Discover the Stars

ALSO BY RICHARD BERRY

Build Your Own Telescope

Discover the Stars

RICHARD BERRY

Harmony Books *New York*

Published by Harmony Books, a division of Crown Publishers, Inc., 225 Park Avenue South, New York, New York 10003, and represented in Canada by the Canadian MANDA Group
HARMONY and colphon are trademarks of Crown Publishers, Inc.
Manufactured in the United States of America

Design by Janet Tingey
Horizon silhouettes by Patricia LaBreque
Sky maps, star charts, and additional line art by Richard Berry

Library of Congress Cataloging-in-Publication Data
Berry, Richard, 1946–
Discover the stars.
Bibliography: p.
1. Astronomy—Observers' manuals. I. Title.
QB63.B476 1987 523 86-27031
ISBN 0-517-56529-3

10 9 8 7 6 5 4 3 2 1
First Edition

Contents

Introduction

The book you are now holding was written to help you discover the stars. It will teach you how the sky moves and help you to appreciate the many things that can be learned from the stars. And it will make sure you have fun as you learn.

I've divided this book into two main sections. The first part focuses on twelve round sky maps that show the entire sky for each month of the year. You don't need a telescope at all to use these maps. They will help you find your way around the sky and learn the constellations.

The second main section comprises twenty-three rectangular star charts that show the constellations in more detail. These charts invite you to search out and see for yourself the interesting stars and the deep-sky wonders tucked among them—with the aid of no more than a binocular or a small telescope.

My advice for using this book is first to read the whole book through, to get an idea of all it can show you. Then come back to chapter 1, "Exploring the Night Sky," to be sure you understand all the basics it explains. Next, read the monthly sky map for the month you're in. Then you'll be ready to take this book outside and start observing.

Once you've learned your way around the sky that month, you can investigate the star charts in chapter 3, which show smaller areas of the sky in more detail, and start using a binocular or telescope to explore even more of the sky. Then as time passes you'll be able to repeat this process anew each month.

Following the sky maps and star charts is a collection of short essays designed to help you explore the moon and planets, and to help you move into more complex areas of star watching.

As you explore the sky with *Discover the Stars*, there's little doubt that your interest in astronomy will grow. Soon you will want to learn more about the celestial objects you have seen, or to read up on the latest findings from satellite observatories. To guide you beyond this book, I have included a listing of star atlases, books, and magazines. You will soon find dozens of new paths to follow as you continue to discover the stars.

1. Exploring the Night Sky

It is a pity, in an age of rockets and space telescopes, that so few people have a direct acquaintance with the stars. Learning the stars and following their nightly courses across the sky brings a deep satisfaction, a satisfaction born of familiarity with something both ancient and ageless.

The first major change you'll notice your first night out watching the stars is that they move across the sky, mostly from east to west. This is because the earth is rotating on its axis from west to east, giving the stars, like the sun and the moon, the appearance of movement in the opposite direction.

Another major change you'll notice over a period of months is that entirely different groups of stars become visible. This is because the earth is circling the sun and because each day the sun's light blocks the stars in the daytime sky from our view. If the sun gave no light, we would be able to see nearly all the stars in the sky every twenty-four hours as the earth completes a full rotation on its axis. As it is, we can see about three-quarters of the stars on any given night, because if we stay up all night, new stars come into view.

As the earth revolves around the sun, different sections of the universe are gradually revealed to us. The part of the universe we view at night in December is mostly invisible to us on nights in June. This is because the earth is on the opposite side of the sun in June from where it is in December, so at night the opposite half of the universe is revealed to us.

As you look at the night sky, you'll notice that groups of stars are easier to recognize than individual stars. People seem always to have known this, for since the beginning of time they have given names to distinctive groups of stars and made up stories about them. These groups of stars that form pictures are called *constellations*.

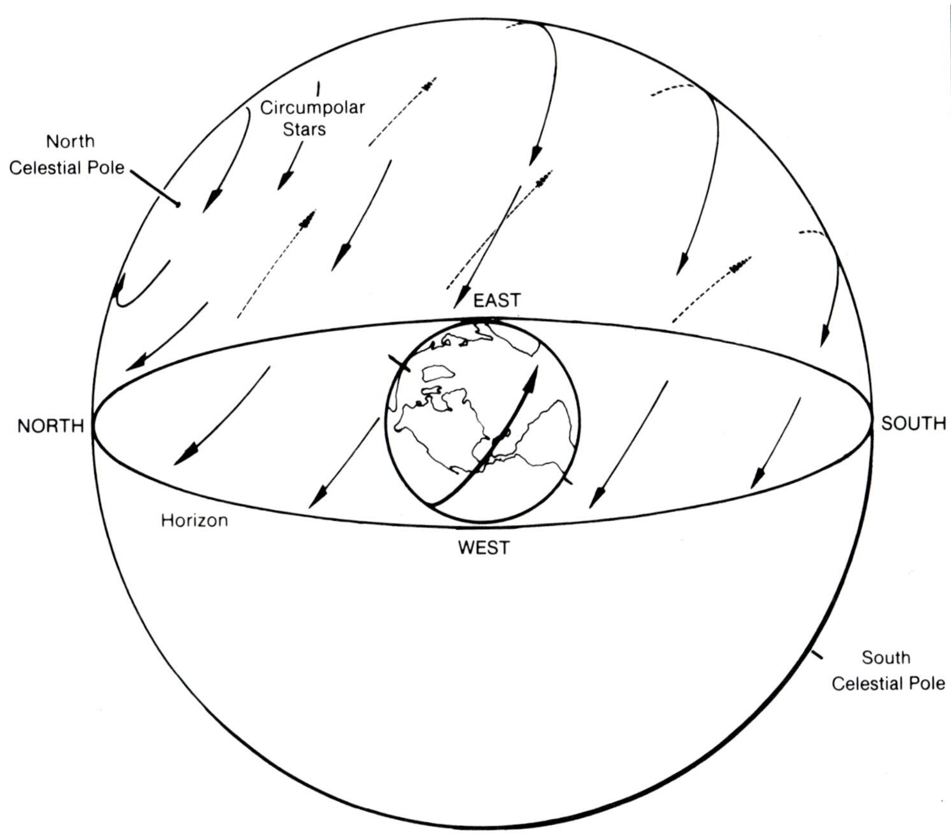

The stars appear to move from east to west.

The constellations are named for various animals, mythic heroes, birds, and monsters. Thus we see Leo, the Lion; Orion, the Hunter; Cygnus, the Swan; and Draco, the Dragon, in our skies. All are simple line or stick figures. Some resemble their namesakes closely, others do not.

Over forty constellations date back at least 2,200 years, to the time of the Greek astronomer Hipparchus, and were listed in his great star catalog. Those constellations were derived from even more ancient Chaldean and Babylonian star groups more than 4,000 years old. Although the star catalog of Hipparchus is lost, a catalog compiled in A.D. 150 has survived. The forty-eight constellations listed in it form the basis for our modern ones. Arab scholars and astronomers preserved the ancient Greek scholarship through Europe's Dark Ages, and many (if not most) stars acquired Arabic names.

When astronomers created new star catalogs during the Renaissance, they often made up new constellations. In 1603, for example, Johann Bayer added twelve new constellations in the *Uranometria,* his star atlas. Subsequent sky mappers—astronomers Johannes Hevelius, John Flamsteed, and Nicolas de LaCaille—squeezed their own constellations between the ancient ones and invented new constellations to fill the southern sky. We now recognize eighty-eight constellations agreed upon by an international body of astronomers in 1930. A list of these constellations appears in the List of Constellations on pages 116–117 at the end of the book.

There are 88 constellations, and about two-thirds of them are visible in the continental United States.

The round star maps that follow are your guide to the bright stars and constellations in the sky. They are suitable for most of North America and any place in the world between latitude 30° and 45° north—from Maine to Texas, Oregon to Louisiana. Outside this range, stars in the north and south will appear a bit higher or lower in the sky, but stars overhead remain essentially correct.

The maps include enough stars to allow you to identify all the major constellations. The brightest stars are labeled in upper- and lowercase letters: for example, Rigel, Regulus, or Antares. The names of the major constellations are shown in large uppercase letters: for example, ORION, LEO, SCORPIUS. Some of the less-conspicuous constellations are named with small uppercase letters: CANCER, CRATER, PISCES.

Each monthly sky map is a picture showing the entire sky as it would appear to you if you were lying flat on your back looking straight up. The circular edge of the map is the horizon, the limit of your vision. The center of the map is the zenith, the point directly overhead in the sky.

Around the edge of the map, you will see the letters *N, E, W,* and S. These show the compass directions north, east, west, and south. If you stand facing south, hold the map with the S at the bottom, just as it's printed in the book. Between the horizon and zenith you will see the stars shown on the map between the bottom edge of the map and its center. If you stand facing in a different direction, turn the book so that direction is at the bottom of the map. The map will then show the stars visible in that direction between its bottom edge and the center.

The *north celestial pole*, which is the point in the sky around which all other stars appear to circle, is labeled "ncp." The sky maps also show a thin line running east-west, called *ecliptic*. This is the path of the sun through the sky. The ecliptic derives its name from the fact that eclipses occur on it. If you see a bright "star" that's not shown on the sky map, and it's near the ecliptic, the "star" is probably a planet. Chapter 5 will help you identify which planet you see.

If you are observing at home, a map of your town or city may help you determine which way is north, south, east, and west. City streets and country roads are often laid out in a north-south and east-west grid. If you are using the maps in a place unfamiliar to you, a compass can help you orient yourself; yet because the magnetic north pole does not coincide with the true north pole, a magnetic compass points north only approximately. The sun also is a rough guide to direction. It rises in the east and sets in the west, but the precise direction of sunrise and sunset varies with the seasons. The best way to establish directions is from the stars themselves. The sky maps will teach you to locate the stars of the Big Dipper, Little Dipper, and Cassiopeia, and from them you can identify Polaris, the polestar or North Star. When you face Polaris, west is to your left and east to your right. South lies directly behind you. If you consistently do your stargazing from one spot, note objects on the horizon which correspond to the cardinal directions, or mark these directions with wood stakes painted white.

Stars appear to move quite differently depending on which part of the sky they're in. Stars low in the east early at night will pass high overhead during the course of the night and set in the west. Stars in the north appear to move in a small circle around the north celestial pole. (They are called *circumpolar stars*.) And stars that rise low in the southeast will move slowly over the southern horizon, reach their high point when due south, then set in the southwest. This movement may sound complicated, but after a few nights out under the stars you will find it easier to picture the earth as a planet moving through space, and understand the logic of these movements.

Because it will be dark outside, you want to use a flashlight to illuminate the map you're using. But an ordinary flashlight is much too bright, since it takes several minutes for your night vision to return

The Milky Way is made up of thousands of faint stars in our galaxy. It appears on the sky maps as a light path running across the sky.

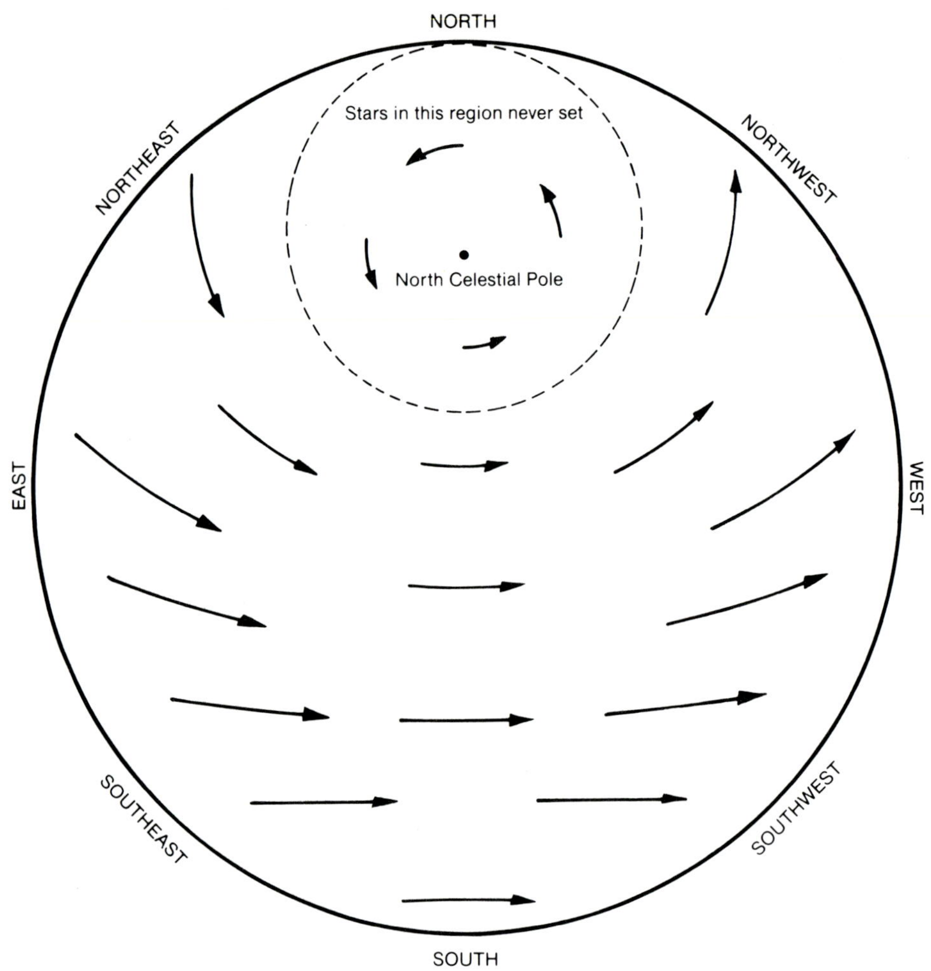

How stars move on the star maps. The arrows show how far a star will move in two hours. Stars in the northern sky circle the north celestial pole. In the rest of the sky, stars rise in the eastern sky, move across the sky, and set in the western sky.

after seeing a bright light. So you'll need to make an astronomer's flashlight. To do this, place two layers of brown paper bag paper inside the lens of an ordinary two-cell flashlight. You can also add a cardboard shield around the end of the flashlight to keep the light from glaring into your eyes.

When you go out at night to look at the stars, go to the largest clearing you can conveniently get to. The larger the clearing, the more sky you'll see. Take a blanket to sit on, and warm clothes—you'll be surprised how cold it can get in the dead of night, even during summer. The farther from a large city you are, the better, because the lights and haze of a city hide many stars. Use the table

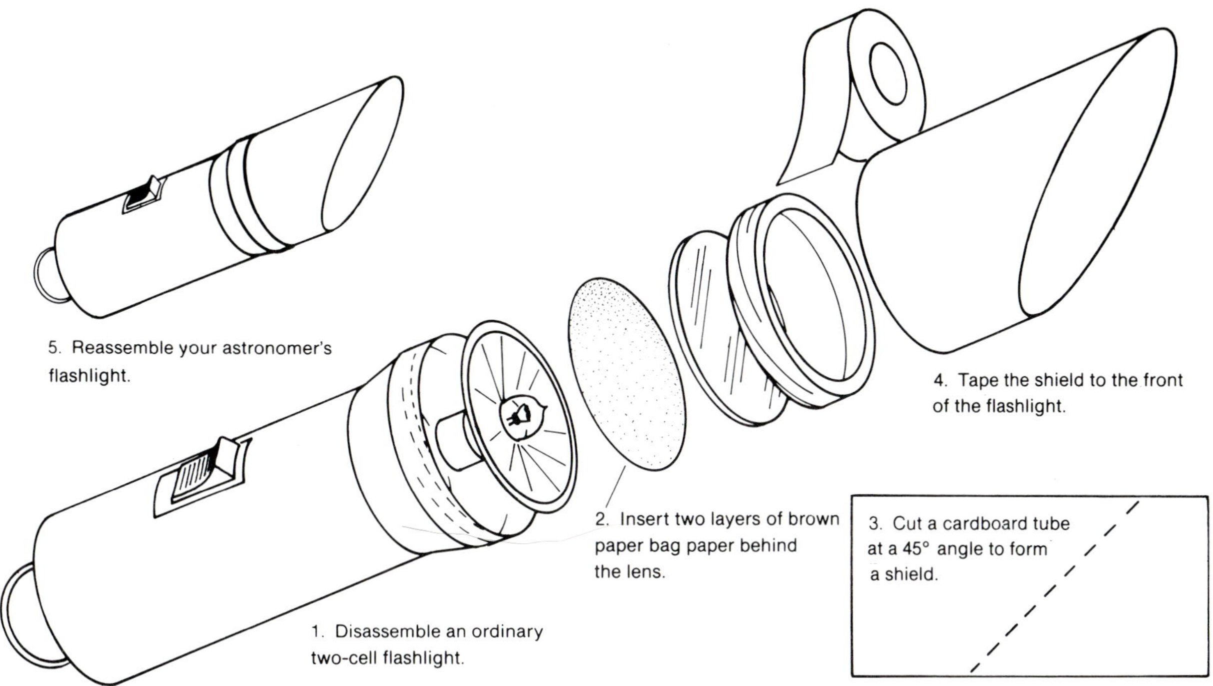

How to make an astronomer's flashlight

at the end of this chapter to help find which sky map is right for the date and time when you're out, orient the chart using your astronomer's flashlight, and start looking. You'll be amazed at how much there is to discover.

The sky maps that follow show the entire sky overhead as it appears at 10:00 P.M. on the seventh day of each month. Each map also shows how the sky will look in other months, at different times of the night. The map called "The March Sky" is accurate on March 7 at 10:00 P.M., but also on February 7 at midnight, on January 7 at 2:00 A.M., and so on. (Think about the 24 hours in the day and the earth's movement through space during the 12 months of the year and this two-hour difference per month will make sense.)

To find the right sky map for the date and time you are out observing, look at the following table, "Which Sky Map to Use." Find the nearest date in the left-hand column and the nearest time in the top row, then read off the name of the appropriate map. This table is repeated on page 118 at the end of the book to make it easier to find when you're out in the dark.

Which Sky Map to Use

	6 P.M.	7 P.M.	8 P.M.	9 P.M.	10 P.M.	11 P.M.	MID-NIGHT	1 A.M.	2 A.M.	3 A.M.	4 A.M.	5 A.M.	6 A.M.
Jan. 7	Nov		Dec		Jan		Feb		Mar		Apr		May
Jan. 23		Dec		Jan		Feb		Mar		Apr		May	
Feb. 7	Dec		Jan		Feb		Mar		Apr		May		Jun
Feb. 21		Jan		Feb		Mar		Apr		May		Jun	
Mar. 7	Jan		Feb		Mar		Apr		May		Jun		Jul
Mar. 23		Feb		Mar		Apr		May		Jun		Jul	
Apr. 7		Feb		Mar		Apr		May		Jun		Jul	
Apr. 22	Feb		Mar		Apr		May		Jun		Jul		Aug
May 7		Mar		Apr		May		Jun		Jul		Aug	
May 23	Mar		Apr		May		Jun		Jul		Aug		Sep
Jun. 7		Apr		May		Jun		Jul		Aug		Sep	
Jun. 23	Apr		May		Jun		Jul		Aug		Sep		Oct
Jul. 7		May		Jun		Jul		Aug		Sep		Oct	
Jul. 23	May		Jun		Jul		Aug		Sep		Oct		Nov
Aug. 7		Jun		Jul		Aug		Sep		Oct		Nov	
Aug. 23	Jun		Jul		Aug		Sep		Oct		Nov		Dec
Sep. 7		Jul		Aug		Sep		Oct		Nov		Dec	
Sep. 22	Jul		Aug		Sep		Oct		Nov		Dec		Jan
Oct. 7		Aug		Sep		Oct		Nov		Dec		Jan	
Oct. 23	Aug		Sep		Oct		Nov		Dec		Jan		Feb
Nov. 7	Sep		Oct		Nov		Dec		Jan		Feb		Mar
Nov. 22		Oct		Nov		Dec		Jan		Feb		Mar	
Dec. 7	Oct		Nov		Dec		Jan		Feb		Mar		Apr
Dec. 23		Nov		Dec		Jan		Feb		Mar		Apr	

This table shows you which sky map to use on any night of the year, at any time of night. For instance, if you're outside on August 7 at 9:00, you should use the map for the July sky. If you stay out the same night until 10:00 you can use the map for either the July sky or the August sky, depending on whether you're more interested in the stars that are setting in the west or the ones that are rising in the east. If you stay out until 11:00, the stars shown on the August sky map will fill the sky.

On this table, and on all the smaller tables facing the individual sky maps, times for January, February, March, November, and December are given in standard time. Times for April, May, June, July, August, September, and October are given in daylight savings time.

When you want to study a particular constellation in detail, turn to the List of Constellations on pages 116–117 at the end of the book. This will tell you which of the monthly sky maps features it most clearly and which of the star charts in chapter 3 it appears in.

The magnitudes (or relative brightness) of the stars on the sky maps are shown in the figure at right. For more on magnitudes, see the explanation on page 46.

Magnitude Scale for Sky Maps

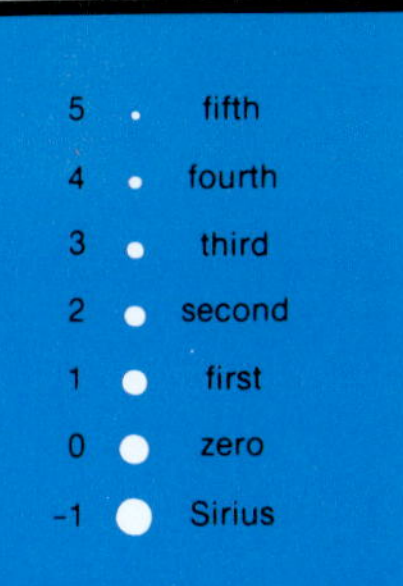

THE JANUARY SKY

The winter stars shine their brightest on crisp January evenings. It is then that the brilliant constellations Orion, Taurus, Gemini, and Auriga ride highest in the sky. Winter's Milky Way, though not as bright as the Milky Way of summer, spans the sky in starry splendor. (See chapter 3 for more on the Milky Way.)

Orion is the most glorious of all the constellations. Even if you have never seen a constellation before, you can spot the three bright stars in a row that make up Orion's belt. Surrounding the belt is a prominent rectangle of bright stars. Above the belt, to the east, bright Betelgeuse (pronounced "beetle-juice") and Bellatrix mark Orion's shoulders.

Orion, the mighty hunter, sword slung from his belt, stands one arm uplifted with a club. The other arm holds a shield against Taurus, the charging bull. When he was stung fatally by the scorpion, Orion was placed among the stars by the gods. When Scorpius rises, Orion sets; when Orion rises, Scorpius sets. They never appear together in the sky.

Mighty Orion helps stargazers; his stars point the way to other constellations. Follow the line of the belt stars to the west to Taurus, the Bull, charging Orion. The bull's face is a V of stars called the Hyades. The brightest star, the bull's eye, is Aldebaran. The tips of his long horns are marked by stars; his body is the Pleiades star cluster.

Taurus, the Bull, is incomplete. Although a number of faint stars are sometimes co-opted to form his legs, you need only the lowered head, horns, and the hint of a body behind to imagine the entire bull. Likewise Orion, though equipped with a sword, club, and shield, lacks legs below his knees, and has no head. The powerful torso and arms alone evoke the hunter.

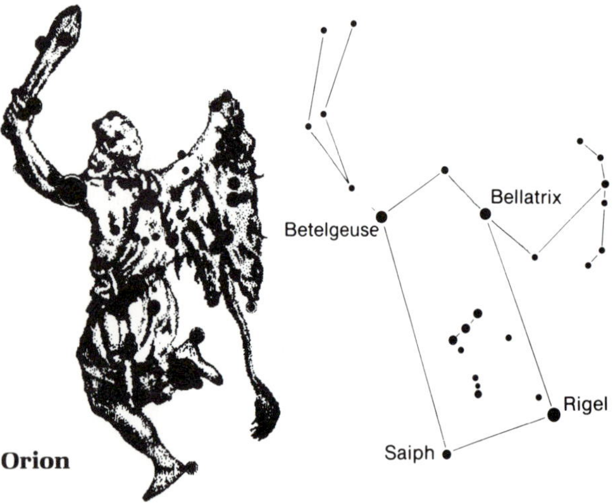

Orion

North of Orion is Auriga, the Charioteer. This constellation is more easily seen as a geometric figure; the distinctive six-sided form is easy to recognize and remember. Oddly enough, Auriga is associated with goats; indeed, the name of the brightest star, Capella, means "little goat," and two stars just south of Capella are called "the kids." On January nights, Auriga is straight overhead during the late evening.

Follow Orion's belt away from Taurus to the dog star, Sirius. Sirius is the eye of Orion's hunting companion, Canis Major, the Greater Dog, and the brightest star in the sky. Orion also points to Gemini, the Twins. From Rigel, trace a line to Betelgeuse until you reach Castor and Pollux, the Heavenly Twins.

The bright stars of the winter constellations form an enormous circle in the sky. Although not a constellation, the Winter Circle will help you recognize the stars and constellations that make it up. Starting at Rigel, trace this giant arc through Aldebaran, Capella, Castor, Pollux, Procyon, Sirius, then back to Rigel. The Winter Circle is so large that it's difficult to take it all in at once. Betelgeuse lies near its center.

As the Winter Circle reaches its highest, Cetus (the Whale), Pisces (the Fishes), and Eridanus (the River) have begun to set. The Great Square of Pegasus, which dominates the autumn sky, sinks in the west, followed by Andromeda and her mythic rescuer Perseus. Cassiopeia and Cepheus lie west of Polaris, and Ursa Minor (the Little Bear) hangs under the polestar.

Yet even as winter's stars shine in their fullest glory, spring is not far behind. In the east, bright Regulus and the sickle of Leo have risen, and the Great Bear noses his way into the northeast sky.

This map shows the sky as it appears on:

October 7 at 5 A.M.; October 23 at 4 A.M.; November 7 at 2 A.M.; November 23 at 1 A.M.; December 7 at midnight; December 23 at 11 P.M.; January 7 at 10 P.M.; January 23 at 9 P.M.; February 7 at 8 P.M.; February 23 at 7 P.M.

NORTH
NORTHEAST
NORTHWEST
WEST
SOUTHWEST
SOUTH
SOUTHEAST
DRACO
URSA MINOR
NCP
Polaris
URSA MAJOR
Dubhe
CEPHEUS
CASSIOPEIA
Caph
CAMELOPARDALIS
ANDROMEDA
PEGASUS
LEO
Regulus
PERSEUS
Algenib
AURIGA
Capella
Algol
TRIANGULUM
CANCER
Castor
Pollux
GEMINI
The Pleiades
Hamal
ARIES
PISCES
ECLIPTIC
TAURUS
Aldebaran
CANIS MINOR
Procyon
HYDRA
Betelgeuse
Alrischa
Mira
ORION
CETUS
MONOCEROS
Rigel
Sirius
LEPUS
ERIDANUS
CANIS MAJOR
FORNAX
PUPPIS
COLUMBA

THE FEBRUARY SKY

Orion dominates the February sky much as he does the January sky—but this most recognizable of constellations now yields the meridian to Canis Major and Canis Minor, his greater and lesser dogs. High overhead are the bright zodiacal twins called Gemini. These three figures comprise the eastern half of the great Winter Circle of bright stars. Its western side is made up of the bright stars in Orion, the Hunter; Taurus, the Bull; and Auriga, the Charioteer.

Canis Major dogs the heels of Orion. Its brightest star, indeed the brightest star in the entire sky, is Sirius, the dog star. To find Sirius, continue the line of Orion's belt stars toward the south and west. You won't miss Sirius sparkling with bluish white light.

Sirius has long figured in both myth and science. To the ancient Egyptians Sirius signaled the flooding of the Nile. They watched the sky just before dawn, and when they saw Sirius rising just before the brightening sky blotted it out, they knew that soon their fields would be renewed again. Likewise, our term "dog days" refers to the hot days of August when Sirius rises with the dawn. In modern times, Sirius has gained fame as a star only eight light-years from our sun, and for the faint white-dwarf companion star, affectionately known as the Pup, that circles it every fifty years.

Sirius represents the eye of Orion's faithful companion. A triangle of stars south of Sirius comprises the dog's hindquarters and tail; his muzzle is the star preceding Sirius across the sky. But Orion has two dogs. While the big dog follows close on the heels of Orion, the little dog runs freely behind him. Canis Minor has only two stars. The brighter of the two is Procyon, a star in the Winter Circle. Procyon means "before the dog" because it rises a few minutes sooner than Sirius. As they race across the sky, though, the little dog follows the larger, brighter dog.

Overhead, the twin stars Castor and Pollux cross the meridian. These stars are the twins of the zodiacal constellation Gemini. Two parallel rows of stars mark their bodies; a perpendicular row marks their joined arms. The Castor and Pollux of Greco-Roman myth were the sons of Jupiter and Leda who sailed with Jason's Argonauts in search of the Golden Fleece.

The ship *Argo*, too, became a constellation. Although Argo is no longer one constellation but four, its high poopdeck, Puppis, lies below Canis Major in the south. Besides Argo's poopdeck, some stars in Vela, the Sails of Argo, peek above the southern horizon in February and March skies.

South of the zenith the February sky is filled with bright stars, but to the north the sky is dark and starless. Ancient astronomers seem to have felt no compulsion to fill this dark region, but celestial cartographer Johannes Hevelius, in the seventeenth century, did. Lynx, the celestial Lynx, and Camelopardalis, the Giraffe, now fill this apparent void. Locate these dim star groups on star chart 4.

In the southwest, the last stars of the "watery" autumn constellations move steadily toward the horizon. In the southwest Cetus, the Whale, is half set, and the twists and turns of the great River Eridanus lie in the trees. In the northwest, Aries, the Ram, is setting, as are the figures of the epic story of Perseus and Andromeda.

All the while, the spring constellations Leo and Ursa Major are already rising in the east. It's easy to spot the little group of stars marking the head of the Water Snake, Hydra. They follow just behind Canis Minor. Sharp-eyed observers will see the dim starry spray of Coma Berenices.

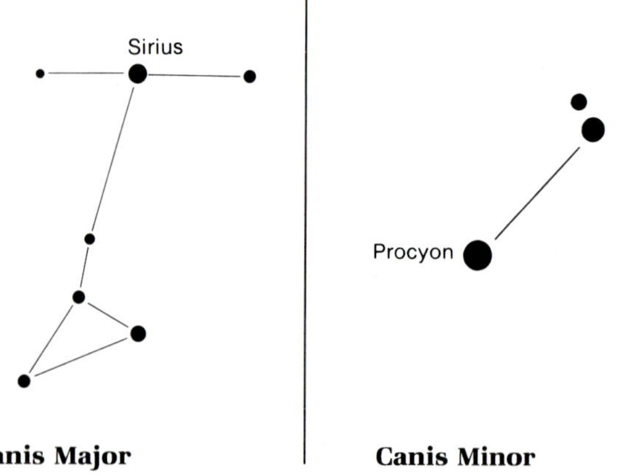

Canis Major **Canis Minor**

This map shows the sky as it appears on:

November 7 at 4 A.M.; November 23 at 3 A.M.; December 7 at 2 A.M.; December 23 at 1 A.M.; January 7 at midnight; January 23 at 11 P.M.; February 7 at 10 P.M.; February 21 at 9 P.M.; March 7 at 8 P.M.; March 23 at 7 P.M.

NORTH

NORTHEAST

NORTHWEST

DRACO

URSA MINOR

CEPHEUS

NCP

Polaris

CASSIOPEIA

Caph

Mizar

ANDROMEDA

CANES VENATICI

Dubhe

CAMELOPARDALIS

URSA MAJOR

Algenib

TRIANGULUM

Algol

COMA BERENICES

PISCES

Capella

PERSEUS

ARIES

AURIGA

The Pleiades

Denebola

Castor

TAURUS

ECLIPTIC

LEO

GEMINI

WEST

Pollux

CANCER

Aldebaran

ECLIPTIC

Regulus

CANIS MINOR

CETUS

Betelgeuse

ORION

CRATER

Procyon

HYDRA

Alphard

MONOCEROS

Rigel

ERIDANUS

Sirius

CANIS MAJOR

LEPUS

PYXIS

PUPPIS

COLUMBA

SOUTHWEST

SOUTHEAST

VELA

SOUTH

THE MARCH SKY

As the brilliant constellations of the winter—Taurus, Orion, Gemini, Auriga, and Canis Major—sink in the western sky, the constellations of spring ascend in the east.

Leo, the Lion, the herald of spring skies, recognizable by the "sickle" asterism that marks his mane and the right triangle of his haunches, glides toward the meridian high overhead. In the northeast, Ursa Major, the Great Bear, which includes the asterism known as the Big Dipper, has risen high in the sky. Use the pointer stars in the cup end of the Dipper to find your way to Polaris, the North Star. Following the arc of the Big Dipper's handle leads you to the bright star Arcturus in Boötes, rising in the west, then to Spica in Virgo, and finally to Corvus, the Crow, still low in the southwestern sky.

Between these bright seasonal groupings lie a few constellations that are less conspicuous, yet date back to the furthest antiquity. Midway between the bright twin stars Castor and Pollux in Gemini and the bright star that marks the heart of Leo, Regulus, lies the dim zodiacal constellation Cancer, the Crab. Two dim stars mark his shell; another two the tips of his extended claws. Cancer faces Leo. The Crab's shell is a tiny trapezoid of dim stars, and within the shell lies the Beehive star cluster. Visible to the naked eye as a faint glow, the Beehive is a pleasing sight through binoculars and small telescopes.

Cancer lies on the ecliptic, the apparent annual path of the sun in the sky. Because the orbits of the planets all lie close to the ecliptic plane in space, if you see a bright "star" in this part of the sky, you are almost certainly viewing one of the planets. Because of the "beastly" nature of many constellations on the ecliptic, they are collectively called the zodiac, from the Greek word for animal. Our word "zoo" shares the same root.

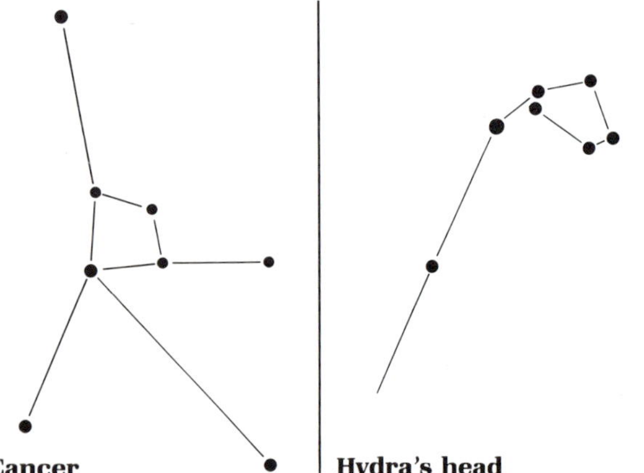
Cancer Hydra's head

Below Cancer, between Procyon and Regulus, lies the head of the longest and largest constellation in the sky—Hydra, the Water Snake. Five stars form the serpent's head. Behind the head, trailing toward the southeastern horizon, is the long body. The brightest star in Hydra is Alphard, the star third from the head.

Follow Hydra's coiling form toward Corvus, the Crow, low in the southeast, to yet another dim constellation, Crater, the Goblet, perched on the snake's back. The snake, the goblet, and the crow figure in a Greek legend. Apollo, the god of the sun, is said to have sent his messenger, the crow, to fetch a goblet of water. On the way the bird spied a fig tree and waited for the figs to ripen. When the crow returned, he carried a water snake in his talons—and told Apollo that the snake had caused the delay. To punish him for this lie, Apollo banished the crow, placing him within sight, but not reach, of water.

Behind Canis Major is Puppis, the Poopdeck of the ship Argo. Jason sailed *Argo* to search for the Golden Fleece. Argo sails backward, the ship's poop (stern) rising first. Although once a major southern constellation, Argo no longer exists. Its stars were divided into more manageable sections: Carina, the Keel; Vela, the Sails; Pyxis, the Compass; and Puppis, the Poop, the high stern of a sailing ship. Vela is still low in the southern sky at its highest, and only part of Carina is visible from the southernmost states.

In the northern sky, Cassiopeia sets very slowly in the northwest. From the southern states, some of its stars will set within a few hours. For northern observers, Cassiopeia will swing under the polestar as the night passes, and before dawn will rise again in the northeast.

EAST

This map shows the sky as it appears on:

November 23 at 5 A.M.; December 7 at 4 A.M.; December 23 at 3 A.M.; January 7 at 2 A.M.; January 23 at 1 A.M.; February 7 at midnight; February 21 at 11 P.M.; March 7 at 10 P.M.; March 23 at 9 P.M.; April 7 at 9 P.M.

NORTH
NORTHEAST
NORTHWEST
CEPHEUS
DRACO
CASSIOPEIA
URSA MINOR
NCP+
Polaris
CORONA BOREALIS
CAMELOPARDALIS
Algol
PERSEUS
BOÖTES
Mizar
Dubhe
Capella
The Pleiades
CANES VENATICI
URSA MAJOR
ECLIPTIC
AURIGA
Cor Caroli
Arcturus
TAURUS
COMA BERENICES
Castor
Aldebaran
WEST
Pollux
GEMINI
VIRGO
Denebola
LEO
CANCER
Betelgeuse
CANIS MINOR
ORION
ERIDANUS
Regulus
Procyon
ECLIPTIC
Rigel
Spica
CRATER
MONOCEROS
HYDRA
Alphard
Sirius
LEPUS
CORVUS
CANIS MAJOR
PYXIS
PUPPIS
SOUTHEAST
SOUTHWEST
VELA
SOUTH

THE APRIL SKY

As the stars of winter set in the west, the constellations of spring rise high. Overhead stalk two powerful beasts, the Great Bear and the Lion, while in the south, the Crow and Goblet ride the back of the twisting Serpent.

Leo (Latin name for lion) reposes with his head high and hind legs tucked under. The lion's head and mane are made up of an arc of stars called the "sickle," a reversed question mark in the sky; the bright star Regulus marks his heart. Three stars form the lion's hindquarters. The easternmost of these stars bears the name Denebola, and marks the base of his tail.

North of Leo, almost straight overhead, is Ursa Major, the Great Bear. The bear's seven brightest stars form a famous asterism, the Big Dipper, so well known that it's almost a constellation within a constellation. Four stars make up the bowl; three stars form the handle. If you face north at this time of year, the Dipper hangs upside down in the sky—but if you lie on your back with your feet to the southern horizon, the bear is upright.

But where is that bear? Locate the bowl of the Dipper, then look on the side opposite the handle. The triangle of faint stars is his head and snout; one star plus two close together make his front leg and forepaw. South of the Dipper's bowl a line of stars ending with two close together make his hind leg and paw. The bowl of the Dipper is the bear's body; the handle is his long tail. Although real bears don't have long tails, celestial bears do. It seems that when the bear was lifted into the sky to be placed among the stars, he was picked up by his tail and it stretched.

The two end stars of the Dipper's bowl are known as the pointers. If you draw a line from Merak to Dubhe and extend it five times its length, you come to Polaris, the polestar, the tip of the Little Bear's tail. The Dipper's handle also shows the way to other constellations. Extend the handle's arc about one and a half times its own length, to Arcturus in Boötes, the Herdsman. Continue this arc to Spica, the brightest star of Virgo, the Virgin. Spica marks the ripe head of a wheat stalk, long a symbol of fertility. You can find Boötes and Virgo with this jingle:

Follow the Arc to Arcturus,
and Speed to Spica—
then Curve to Corvus!

Curve to Corvus? Corvus is a stellar trapezoid just below Virgo, and the brightest constellation in this rather dim sky region. Corvus, the Crow, makes sense when you make a cross of his stars—a bird's body and wings. Though we regard Corvus as a crow, Corvus has also been raven. Both are large, black birds.

Beside Corvus lies a dim constellation, Crater, the goblet of the god Apollo. *Crater* is Latin for "bowl." The same word applies to the moon's craters, bowl-like formations that pock the lunar surface. Twining past Corvus and Crater is the body of Hydra, the Water Snake. Hydra's head precedes Leo across the sky; though faint, Hydra is the longest of the constellations, spanning nearly one-third of the sky.

In spring the Milky Way is at its least conspicuous. The north pole of the Galaxy lies in Coma Berenices, so when this constellation reaches its apex, the Milky Way lies on the horizon, lost in haze and light pollution. Yet there is compensation, for among the relatively dim constellations of spring lie thousands of galaxies, far-off milky ways to other suns like ours. When we gaze up to the spring stars, we are also looking out of our galaxy to the galaxies beyond. With a telescope you can see these "island universes" as faint glows nestled among the foreground stars of our galaxy.

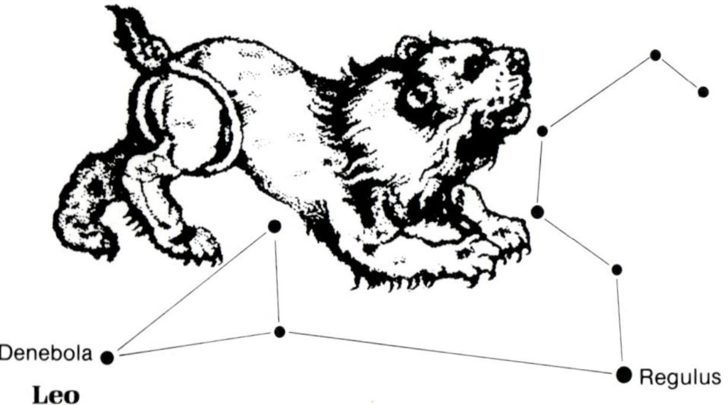

Leo

This map shows the sky as it appears on:

December 23 at 5 A.M.; January 7 at 4 A.M.; January 23 at 3 A.M.; February 7 at 2 A.M.; February 21 at 1 A.M.; March 7 at midnight; March 23 at 11 P.M.; April 7 at 11 P.M.; April 22 at 10 P.M.; May 7 at 9 P.M.

NORTH
NORTHEAST
NORTHWEST
WEST
SOUTHWEST
SOUTH
SOUTHEAST
CEPHEUS
CASSIOPEIA
PERSEUS
Vega
DRACO
Polaris
NCP
CAMELOPARDALIS
URSA
MINOR
Capella
AURIGA
HERCULES
Dubhe
CORONA
BOREALIS
Mizar
URSA MAJOR
ECLIPTIC
TAURUS
CANES
VENATICI
Gemma
Castor
Cor Caroli
GEMINI
Pollux
COMA
BERENICES
BOÖTES
Betelgeuse
ORION
CANCER
Arcturus
CANIS
MINOR
SERPENS
CAPUT
LEO
Denebola
Procyon
Regulus
VIRGO
MONOCEROS
Sirius
CRATER
LIBRA
HYDRA
Alphard
Spica
ECLIPTIC
CANIS MAJOR
CORVUS
PUPPIS
PYXIS
CENTAURUS
VELA

THE MAY SKY

Overhead Arcturus, the bear stalker, tracks Ursa Major on his endless diurnal circuit of the sky, while Leo, the Lion, presses westward. To the east the first summer stars are rising; in the west, the last of the winter stars will soon set. Of the cold weather luminaries, only Castor and Pollux, the Twins, still shine well above the treetops, while bright Procyon and Capella twinkle through their upper branches.

To the southeast, ruddy Antares, heart of the Scorpion, has risen, while in the northeast, blue-white Vega of Lyra, the Lyre, sparkles, and Deneb of Cygnus, the Swan, pokes above the trees. In another hour or so, white Altair will rise to join Vega and Deneb, completing the Summer Triangle. The dark hulk of Ophiuchus, the Serpent Handler, is rising, soon to be followed by the gentle arch of summer's Milky Way.

From Ursa Major, now overhead, follow the pointers of the Big Dipper to Polaris, the polestar. From this star, which always appears at the same place in the northern heavens, the tail and body of Ursa Minor, the Lesser Bear, arc toward the tail of the Great Bear. The line of stars representing Draco, the Dragon, winds between the two bears; Draco's four-starred head lies between the body of the Lesser Bear and Lyra, the Lyre.

Now follow the graceful arc of the Dipper's handle to the bright yellowish star Arcturus, the luminary of Boötes, the Herdsman, then speed on to Spica in the ancient zodiacal constellation Virgo, the Virgin.

Whether in the guise of fertility goddess, saint, or maiden of the planting season, these stars have represented a young woman associated with the arrival of spring as far back in time as we can trace the constellations, and in many different cultures. The very name of the star Spica comes from the Latin word for an ear of wheat, a symbol of agricultural fecundity. Virgo lies on the ecliptic, the path of the sun through the sky, so one or more of the planets often accompany Spica.

Although Virgo is not a particularly bright constellation, Spica is easy to find from the Dipper, so Virgo, too, is easy to find. Starting at Spica trace out the stick-figure form of a reclining woman. Her arms are raised in a circle around her head. As is the case with many of the constellations representing humans, Virgo has arms and legs but no head.

Continuing the arc from Arcturus past Spica, you'll next encounter Corvus, the Crow. From Corvus it's a short hop to Crater, the Goblet, then to Hydra, the Water Snake, spanning the southern sky.

North of Virgo is another constellation of feminine origin: Coma Berenices, the Hair of Berenice. Legend has it that the Egyptian queen Berenice swore to the gods that she would cut off her beautiful hair if her husband returned safely from war. When he returned, honorable Berenice kept her pledge and placed her locks on the altar, an act which so pleased the gods that they placed her hair in the heavens for all to see. The constellation consists of three dim stars set in an upside-down L, and a loose cluster of very faint stars which represent the hair.

Neither Virgo nor Coma Berenices are bright star groups, but they happen to lie near the north pole of our home galaxy, the Milky Way. In this part of the sky, far from the obscuring dust and gas of our own galaxy, light from galaxies beyond our Milky Way reaches us. With a telescope these rather dim constellations are rich in distant "milky ways."

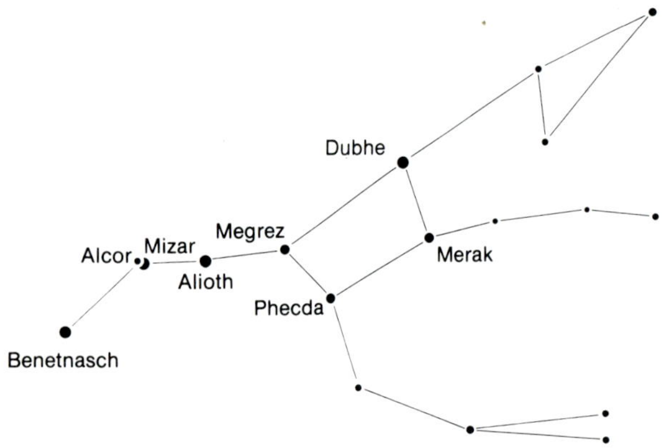

Ursa major

This map shows the sky as it appears on:

January 23 at 5 A.M.; February 7 at 4 A.M.; February 21 at 3 A.M.; March 7 at 2 A.M.; March 23 at 1 A.M.; April 7 at midnight; April 22 at midnight; May 7 at 11 P.M.; May 23 at 10 P.M.; June 7 at 9 P.M.

NORTH
CASSIOPEIA
CEPHEUS
Capella
CAMELOPARDALIS
NORTHWEST
Deneb
Polaris
NCP
AURIGA
CYGNUS
DRACO
URSA
MINOR
GEMINI
URSA
MAJOR
Vega
Castor
Dubhe
LYRA
Pollux
Mizar
HERCULES
ECLIPTIC
CORONA
BOREALIS
CANES
VENATICI
CANCER
CANIS
MINOR
WEST
Cor Caroli
Procyon
Rasalhague
COMA BERENICES
Gemma
LEO
Rasalgethi
BOÖTES
Arcturus
SERPENS
CAPUT
Regulus
Denebola
OPHIUCHUS
VIRGO
Alphard
HYDRA
CRATER
CORVUS
Spica
ECLIPTIC
LIBRA
Antares
SCORPIUS
SOUTHWEST
SOUTHEAST
LUPUS
CENTAURUS
SOUTH

THE JUNE SKY

The Great Bear has passed its highest point but it continues to dominate the heavens, for the stars of Ursa Major point the way to other sky patterns. The two end stars of the dipper point to Polaris, the polestar, while the arc of the handle continues to bright Arcturus in Boötes, the Herdsman, and on to Spica, the bright star in Virgo, the Virgin.

Despite the Bear's continued eminence, the guardians of spring are setting. Leo now slips toward the western horizon while Corvus and Crater sink in the southwest; only the tail of Hydra remains high. Meanwhile, in the east, new stars rise. In the southeast lurks fearsome Scorpius, followed by Sagittarius, the Archer. Spanning the eastern horizon is the Milky Way, bearing two great birds, Aquila, the Eagle, and Cygnus, the Swan, and in the north, Cassiopeia is once again on the rise.

Virgo has crossed the meridian, but remains high in the southwestern sky, every bit as high as Ophiuchus, the Serpent Handler, rides in the southeast. Above Virgo is Coma Berenices, memorial to a woman's beauty, while above Ophiuchus kneels the exhausted figure of Hercules, hero of the dozen labors.

In the north, Ursa Minor, the Lesser Bear, stands straight up on the tip of his tail marked by the polestar, Polaris. Smaller, dimmer, less conspicuous than the showy Great Bear, this Lesser Bear likewise contains a Dipper. The bowl contains four stars, the handle three, curving backward like the handle of a ladle in a bowl of fruit punch. Twisting past Ursa Minor, the coils of Draco, the Dragon, wrap from Hercules to Ursa Major. The Dragon's head lies under the foot of kneeling Hercules.

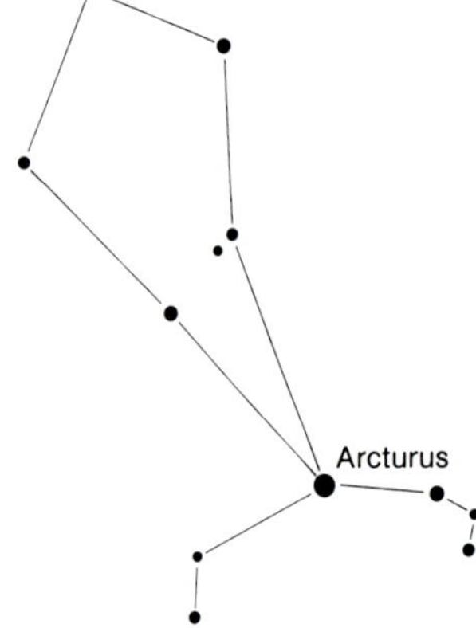

Boötes

Long associated with the Great Bear, the bright star Arcturus in the constellation Boötes (pronounced bow OH tees) stands almost straight overhead. Orange or yellowish in color, Arcturus is the brightest star in the northern sky. Only three stars, all south of the celestial equator (Sirius, Canopus, and Alpha Centauri), are brighter. Boötes represents the figure of a herdsman, coming behind and seeming to urge the Bear endlessly around the pole. Indeed, the name of the star, Arcturus, means "bear-guard."

Boötes resembles an ice cream cone with Arcturus at the lower tip. Two lines of stars form the cone. The ice cream scoop is even completed by a star—the cherry on top, of course! The ancients saw the "cone" as the torso of a man, and may have seen a line of faint stars that tend toward the tail of the bear as the raised arm of the herdsman driving the bear.

A small but conspicuous arc of seven stars called Corona Borealis, the Northern Crown, lies nearly straight overhead. There is no mistaking this constellation tucked between Boötes and Hercules once you have seen it. The crown is said to be that of Ariadne, placed in the heavens after her death.

In the south, kite-shaped Libra, the Scales, approaches the meridian. Although it is one of the zodiacal constellations, Libra dates only to the time of the Romans, for whom it represented the scales of justice. For the Greeks, these stars were the claws of Scorpius. Indeed, the names of the two brightest stars in Libra, Zubenelgenubi and Zubeneschamali, mean the southern claw and the northern claw, respectively. Zubeneschamali is the top star in Libra. The sun's path through the sky—the ecliptic—passes almost exactly through Zubenelgenubi. A bright "star" in the area of Libra is almost certainly one of the planets.

This map shows the sky as it appears on:

February 21 at 5 A.M.; March 7 at 4 A.M.; March 23 at 3 A.M.; April 7 at 3 A.M.; April 22 at 2 A.M.; May 7 at 1 A.M.; May 23 at midnight; June 7 at 11 P.M.; June 22 at 10 P.M.; July 7 at 9 P.M.

NORTH
NORTHEAST
NORTHWEST
WEST
SOUTHWEST
SOUTHEAST
SOUTH
CASSIOPEIA
CAMELOPARDALIS
CEPHEUS
Polaris
NCP
URSA
MINOR
DRACO
URSA
MAJOR
Dubhe
Deneb
CYGNUS
Mizar
ECLIPTIC
Vega
HERCULES
CANES
VENATICI
Albireo
SAGITTA
LYRA
BOÖTES
Cor Caroli
LEO
CORONA
BOREALIS
Regulus
Altair
COMA
BERENICES
Gemma
Arcturus
Denebola
Rasalhague
Rasalgethi
SERPENS
CAUDA
SERPENS
CAPUT
VIRGO
OPHIUCHUS
CRATER
SCUTUM
Zubeneschamali
Spica
Zubenelgenubi
ECLIPTIC
Nunki
LIBRA
CORVUS
Antares
HYDRA
SCORPIUS
SAGITTARIUS
Shaula
LUPUS
CENTAURUS

THE JULY SKY

Summer's stars shine in all their glory shortly after sunset during short midsummer nights. Scorpius, one of the most ancient and venerable of all the constellations, reaches its highest point in the south, while overhead a superhero of Greek mythology kneels, and the constellations of the Summer Triangle move toward greatest ascendancy.

In the northern sky, the Great Bear is on his way down, and in his place rises Draco, the dragon of the northern heavens. The head of the dragon is formed from four third-magnitude stars, and lies midway between Hercules and the bright star Vega, in Lyra. Coiling almost halfway around the pole, Draco's body broadens, turns back on itself, then wraps around the bowl of the Little Dipper, and ends between the pointer stars of the Big Dipper and Polaris.

Straight overhead Hercules kneels when darkness descends on July evenings. The hero of twelve great labors, mythic Hercules was betrayed with a shirt that burned and tortured him. To ease his pain, Jupiter took pity on him and placed him in the heavens. As a constellation, Hercules is not very apparent. The most conspicuous feature is the keystone, a remarkably regular trapezoid. From each corner of the keystone a line of stars extends—these are the legs (north) and arms (south) of the hero. In his hand, Hercules holds a lion's skin, a group of stars just south of Lyra.

Of the giants in the sky the least well known is Ophiuchus, the Serpent Handler. Ophiuchus is a large, dim constellation, and rather hard to see. Furthermore, he is overshadowed by the bright Milky Way constellations nearby. To add insult to injury, there is no major mythology associated with the constellation: it simply looks a lot like a stick-figure tough guy wrestling a serpent with his hands. As you trace the outline of the bulky body, note how few faint stars lie inside it. The "empty" region lends the outline much of its unity.

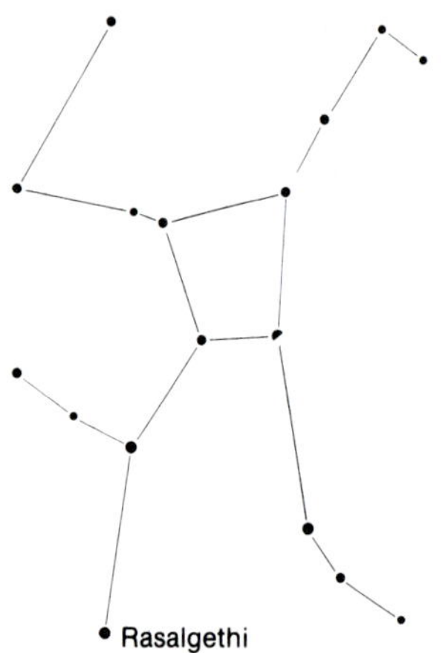

Hercules

The serpent Ophiuchus holds is Serpens. Serpens's head, like the head of Draco, is a triangle of stars; the tail section a faint line of stars bordering the Milky Way. Where the serpent's body crosses the serpent handler the stars belong to Ophiuchus, but in ancient times the two constellations melded together.

Rivaling Orion as the most magnificent of the constellations is Scorpius, the fearsome scorpion of the southern skies. And few figures among the stars so vividly evoke the creature they purport to represent. Scorpius's long curving tail equipped with its deadly barb, the bright star Antares centered in the thorax, the minuscule head, and enormous pinchers all evoke the creature as readily for us today as they once must have for the desert peoples who named these stars.

Incidentally, you have probably heard the Scorpion called "Scorpio." The difference is that the *-ius* ending signifies the astronomical name of the constellation, whereas the *-io* ending is the name of the astrological sign.

Summer's sky is filled with bright stars and conspicuous constellations—far easier to find than spring's dim Crater or sprawling Hydra, now setting, or the vast, dark expanses of the autumn constellations. The reason is that the center of our Milky Way galaxy lies between Scorpius and Sagittarius, the Archer. The Milky Way abounds in bright stars that we use in creating our constellations. Trace this glowing band of light from Scorpius and Sagittarius in the south through Aquila, Cygnus, and Cepheus, to Cassiopeia in the north.

This map shows the sky as it appears on:

April 7 at 5 A.M.; April 22 at 4 A.M.; May 7 at 3 A.M.; May 23 at 2 A.M.; June 7 at 1 A.M.; June 22 at midnight; July 7 at 11 P.M.; July 23 at 10 P.M.

EAST

AQUAF

NORTH

NORTHEAST

NORTHWEST

CAMELOPARDALIS

CASSIOPEIA

URSA MAJOR

Polaris

NCP

URSA MINOR

CEPHEUS

DRACO

Mizar

LEO

CANES VENATICI

COMA BERENICES

Deneb

CYGNUS

HERCULES

Denebola

PEGASUS

Vega

BOÖTES

WEST

CORONA BOREALIS

LYRA

Enif

DELPHINUS

Albireo

Gemma

Arcturus

SAGITTA

EQUULEUS

Altair

Rasalhague

Rasalgethi

SERPENS CAPUT

VIRGO

ECLIPTIC

SERPENS CAUDA

AQUILA

OPHIUCHUS

Spica

Zubeneschamali

SCUTUM

HYDRA

Zubenelgenubi

ECLIPTIC

LIBRA

Nunki

Antares

CAPRICORNUS

SCORPIUS

SAGITTARIUS

CENTAURUS

SOUTHWEST

Shaula

LUPUS

SOUTHEAST

CORONA AUSTRALIS

SOUTH

THE AUGUST SKY

August evenings offer perhaps the finest stargazing there is. The bright stars of Sagittarius, the Archer, grace the southern meridian amid the glowing starclouds of the summer's Milky Way. Ruddy Antares glimmers at the heart of fearsome Scorpius, dominating the southwestern sky. Overhead, the Summer Triangle unites the brilliant Vega, Deneb, and Altair, while the soft shimmer of the Milky Way spans the entire heavens.

In the west the spring stars are setting. Ursa Major, the Great Bear, hangs in the northwest; the handle of the Dipper within the Bear pointing still to Arcturus, the bear-following star of Boötes, the Herdsman. Virgo is gone from the sky, and Libra is soon to follow. Above Boötes, the jewels of the Northern Crown, Corona Borealis, glitter, while Hercules, the doer of deeds, kneels, tired from his labors.

Directly overhead, Vega (pronounced VEE-guh), one of the finest stars in the sky, shines with a steady blue-white light. Vega is the fifth-brightest star in the sky, second brightest in the northern heavens. Compare its light with that of Arcturus, the brightest of the northern stars. Whereas the latter is yellow or orange, Vega's light is the purest blue-white, diamond-hard in its brilliance.

Vega is the *lucida* (brightest star) of Lyra, the Lyre, a small hand-held musical instrument rather like a harp. Four stars make up the body of the Lyre; Vega is a jewel set in it for decoration.

Vega is the brightest of the three stars in the Summer Triangle. Its other stars, Deneb and Altair, belong to the Eagle, Aquila, and Cygnus, the Swan, winging their way south down the Milky Way. Deneb represents the tail of the Swan.

Altair, flanked by a star on either side, is the tail of Aquila, the Eagle. The attendant stars are the bird of prey's talons. Two broad stellar triangles outline the Eagle's wings. Two stars mark his head, outstretched, as he glides southward. Is tiny Sagitta, the Arrow, just behind Aquila, a missed shot intended for the Eagle? Who might have fired an arrow?

Due south, at his highest early on hot August evenings, strides Sagittarius, the Archer. This southernmost of the zodiacal constellations dates back to antiquity. The bow and arrow is a strong and obvious image. Three stars make up the powerful bow; one star tips the arrow and four feather it. By contrast, only a few dim stars mark the archer, insignificant beside his potent weapon.

But Sagittarius reveals a more gentle asterism: the "teapot." The same four stars that feather the arrow are the handle. Five stars outline a dumpy little English pot. A triangle on the west side is the spout. Even the Milky Way gets into the act. Just above the spout the Sagittarius starcloud doubles as a puff of celestial steam.

Yet even as summer's stars shine in their full glory, new stars are rising in the east. In the southeast, the star-tipped horns of Capricornus, the Sea Goat, follow Sagittarius on the ecliptic. Two asterisms, the Circlet of Pisces, the Fishes, and the Water Jar of Aquarius, herald these dim fall constellations. Due east, Pegasus, the Flying Horse, has risen above the trees, his winged body outlined as the Great Square.

The all-star cast of an ancient legend, ofttimes linked to Pegasus, spangle the northeastern sky. Cassiopeia, the vain queen, the foolish king, Cepheus, and their beautiful daughter, Andromeda, chained to a cliff by the sea, have already risen. Only Perseus, her rescuer, has yet to appear on the scene, for some reason trailing far behind his flying steed Pegasus.

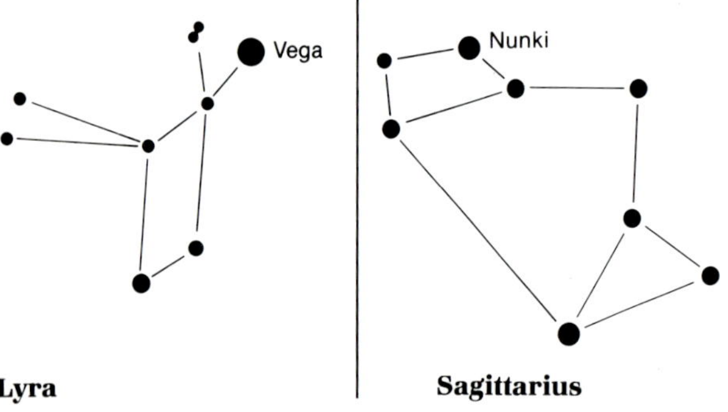

Lyra **Sagittarius**

This map shows the sky as it appears on:

May 23 at 4 A.M.; June 7 at 3 A.M.; June 22 at 2 A.M.; July 7 at 1 A.M.; July 23 at midnight; August 7 at 11 P.M.; August 23 at 10 P.M.; September 7 at 9 P.M.

NORTH
NORTHEAST
NORTHWEST
WEST
SOUTHWEST
SOUTH
SOUTHEAST
CAMELOPARDALIS
PERSEUS
URSA MAJOR
CASSIOPEIA
Polaris
NCP
URSA MINOR
CANES VENATICI
PISCES
CEPHEUS
DRACO
COMA BERENICES
ANDROMEDA
HERCULES
BOÖTES
Deneb
CYGNUS
Vega
CORONA BOREALIS
Arcturus
Gemma
PEGASUS
LYRA
Albireo
DELPHINUS
SAGITTA
SERPENS CAPUT
Enif
Rasalhague
Rasalgethi
VIRGO
Altair
AQUARIUS
EQUULEUS
SERPENS CAUDA
OPHIUCHUS
AQUILA
LIBRA
SCUTUM
ECLIPTIC
Antares
Nunki
Fomalhaut
CAPRICORNUS
SAGITTARIUS
SCORPIUS
Shaula
CORONA AUSTRALIS

THE SEPTEMBER SKY

Seen from dark skies, the Milky Way arches across the heavens. From Sagittarius, the Bowman (Archer), in the southwest, the starry stream flows past Ophiuchus holding the serpent, through Scutum, the Shield, and Aquila, the Eagle, to another bird, Cygnus, the Swan, at the zenith, thence through Cepheus, Cassiopeia, and Perseus rising in the northeastern sky. If your skies are brightened by city lights, make plans to spend a night in a dark place far from the glare of civilization to see the sky as the ancient astronomers—as, indeed, all of humankind—once saw the heavens.

The last spring stars are now setting and the first brilliant stars of winter will not rise for several hours yet. Overhead, Vega, Deneb, and Altair of the Summer Triangle show the way to Lyra, the Lyre, Cygnus, the Swan, and Aquila, the Eagle. Dim constellations—Capricornus, the Sea Goat, Aquarius, the Water Bearer, Pisces, the Fishes, Cetus, the Whale, and Piscis Austrinus, the Southern Fish—"watery" autumnal groups all, occupy the southeastern quadrant of the sky. The best sights lie along the Milky Way.

Straight overhead flies the starry outline of Cygnus, the Swan, neck fully outstretched, winging down the Milky Way. Many ancient peoples seem to have seen this group of stars as a bird, though not necessarily as a swan. Bright Deneb is the tail; Albireo is the head. Five stars wide, the wings sweep back, convincingly swanlike. Within the Swan, Cygnus, is the Northern Cross asterism. A relatively recent invention, the Northern Cross is especially striking on winter nights when the cross, upright, shines in the northwestern sky.

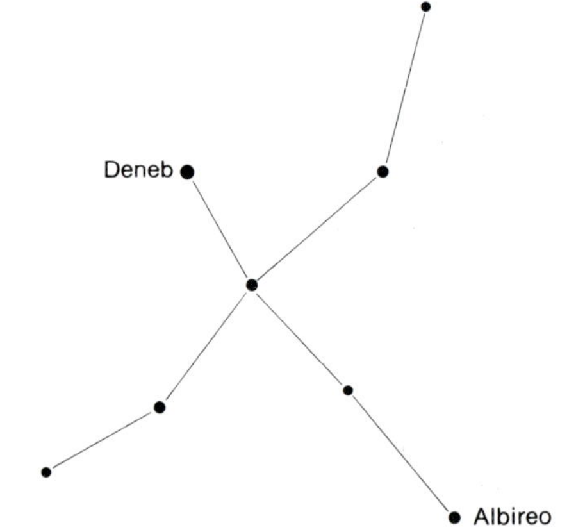

Cygnus

Cygnus lies in the star-rich Milky Way; indeed, some of the starriest views and some of the brightest parts of our galaxy are in the bird's neck. The Milky Way's Great Rift, a dark region that extends from Ophiuchus and through Aquila, ends in the glowing star clouds in northern Cygnus near Deneb. A large but dim cloud of interstellar gas, the North America Nebula, shows just west of Deneb as a brightening in the Milky Way.

Two small ancient constellations, Sagitta and Delphinus, lie just below Cygnus. Sagitta, near Albireo, has long been held to represent an arrow, though exactly which arrow is not clear. Is Sagitta an arrow fired by Hercules? by Sagittarius? Or is the image so strong that Sagitta is an arrow first, and part of legend later? The other small group is Delphinus, the Dolphin, leaping from the waves; the resemblance to the creature is strong.

Following the bright "teapot" of Sagittarius on the ecliptic is the dim sea goat Capricornus. Although they are not very bright, the stars here form a "grin," the remnant of a celestial Cheshire cat. A pair of brighter stars marks each end of the grin. The ancients saw in this shape, perhaps, a goat's horns, or the distinctive outline of a goat's body. The ancient Greeks describe Capricornus as a goat. Because it lies in the "watery" heavens, however, the goat somehow gained a fish's tail and became a sea goat.

As Hercules slowly moves toward the west, Pegasus, the Flying Horse, and Andromeda, the Chained Maiden, approach the meridian from the east. These constellations flank the Milky Way, and lie about the same distance from Cygnus in opposite directions. Don't forget to look for Ursa Major, now low in the northwest, swinging toward its lowest point in the north, the pointers still faithfully indicating Polaris.

This map shows the sky as it appears on:

June 7 at 5 A.M.; June 22 at 4 A.M.; July 7 at 3 A.M.; July 23 at 2 A.M.; August 7 at 1 A.M.; August 23 at midnight; September 7 at 11 P.M.; September 22 at 10 P.M.; October 7 at 9 P.M.; October 23 at 8 P.M.

NORTH
NORTHEAST
NORTHWEST
WEST
SOUTHWEST
SOUTH
SOUTHEAST
URSA MAJOR
CAMELOPARDALIS
URSA MINOR
NCP
Polaris
PERSEUS
CASSIOPEIA
CEPHEUS
DRACO
BOÖTES
HERCULES
Caph
TRIANGULUM
ANDROMEDA
CORONA BOREALIS
Gemma
Hamal
ARIES
Deneb
Vega
CYGNUS
LYRA
SERPENS CAPUT
PISCES
Albireo
SAGITTA
Alrischa
PEGASUS
DELPHINUS
SERPENS CAUDA
Enif
Altair
OPHIUCHUS
EQUULEUS
CETUS
AQUARIUS
AQUILA
SCUTUM
ECLIPTIC
Nunki
Fomalhaut
CAPRICORNUS
SAGITTARIUS
PISCIS AUSTRINUS
CORONA AUSTRALIS
GRUS

THE OCTOBER SKY

As the glories of the summer sky sink in the west and before the brilliant gems of winter rise, autumn's stars grace the meridian. This is not a sky of bright stars, but rather a sea of large, dim figures known as the "watery constellations." Here swim the tied Fishes of the zodiac, Pisces, and the Whale, Cetus. Capricornus, the Sea Goat, dwells here, while Aquarius pours from his Water Jar into the mouth of Piscis Austrinus, the Southern Fish. Bringing up the rear, the River Eridanus rises in the east.

There are still bright stars in the sky, to be sure. The Milky Way arches high; the Summer Triangle is well up in the west. Overhead, the cast of stars in another great mythic drama is near its apex: Andromeda, the beautiful princess; Cassiopeia and Cepheus, the vain queen and cruel king; Cetus, the sea monster who would devour Andromeda; Perseus, her rescuer, and Pegasus, his winged horse. We tell their story in November.

The Great Square of Pegasus, like the Big Dipper in Ursa Major, is an asterism, a constellation within a constellation. Oddly enough, the brightest star in the Great Square doesn't belong to Pegasus; it is officially designated as part of neighboring Andromeda. Pegasus makes a fairly good horse. A line of stars to the south and west suggest the extended neck and head. Two others make the east-reaching legs, and the Great Square is his body. Markab lies at the junction of the body and neck; Markab means "saddle" in Arabic. The legs branch out from Sheat, the shoulder. Enif marks his nose. There is little to suggest the wings of the mythical flying horse, or for that matter, any hind legs, except the star named Algenib, the wing. Originally Pegasus represented the mount of Bellerophon, ridden when he killed the Chimera, but he has since become associated with the Perseus legend.

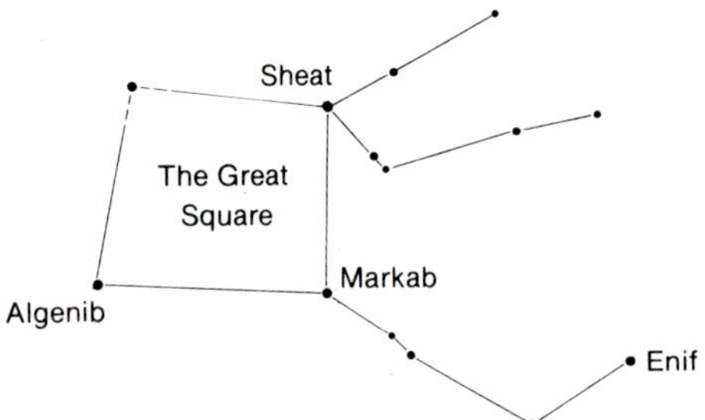

Pegasus

Just south of the head of Pegasus lies the Water Jar, a small, conspicuous asterism in Aquarius, the Water Man. Aquarius is an ancient constellation, but quite dim. Group the stars as a circle above Capricornus for the man, then the Water Jar itself, and an eastern circle marking water from the jar. This eastern circle contains numerous chains of faint stars that suggest flowing or spilling water.

On the meridian low in the south is Fomalhaut, "the mouth of the fish." Fomalhaut is the brightest star in Piscis Austrinus, the Southern Fish. Legend has it that this constellation was an ancient Middle-Eastern fish god variously known as Oannes, Phagre, or Dagon. In any case, it was a fish.

Below and east of the Great Square are more fish; indeed, two fish tied with a cord. The western fish is an asterism called the Circlet of Pisces; the northern fish is a gathering of dim stars. Between them runs a long line of stars in a sharp V, with Alrischa, the knot, at its point. The vernal equinox—where the sun crosses the celestial equator as it moves northward in the spring—now lies in Pisces, just east of the Circlet. When the sun lies in Pisces, of course, you cannot see its stars, but nonetheless, that is where the sun will be when warm weather returns.

In the west, we now see the heroes Hercules and Ophiuchus setting. The birds Aquila and Cygnus are flying toward the horizon. In the north, depending on your latitude, the Big Dipper skims or dips below the horizon; in the south, the Crane, Grus, puts in a brief appearance. And in the east, the stars of winter are waiting in the wings. Already the Pleiades, harbingers of the winter's brilliant stars, are on the rise.

This map shows the sky as it appears on:

July 23 at 4 A.M.; August 7 at 3 A.M.; August 23 at 2 A.M.; September 7 at 1 A.M.; September 22 at midnight; October 7 at 11 P.M.; October 23 at 10 P.M.; November 7 at 9 P.M.; November 23 at 8 P.M.; December 7 at 7 P.M.

NORTH
NORTHEAST
NORTHWEST
URSA MAJOR
URSA MINOR
NCP
Polaris
CAMELOPARDALIS
DRACO
HERCULES
AURIGA
Capella
CEPHEUS
PERSEUS
CASSIOPEIA
TAURUS
Algenib
Algol
Caph
Vega
LYRA
Deneb
The Pleiades
ANDROMEDA
CYGNUS
TRIANGULUM
Albireo
OPHIUCHUS
WEST
ECLIPTIC
Hamal
ARIES
SAGITTA
DELPHINUS
SERPENS CAUDA
PEGASUS
Altair
PISCES
Enif
Alrischa
Mira
EQUULEUS
AQUILA
AQUARIUS
SCUTUM
CETUS
ERIDANUS
ECLIPTIC
Fomalhaut
CAPRICORNUS
PISCIS AUSTRINUS
SOUTHEAST
SOUTHWEST
PHOENIX
GRUS
SOUTH

THE NOVEMBER SKY

In November the stars of summer set while those of winter rise. In the west, Vega, Deneb, and Altair, the triad of bright stars that make the Summer Triangle, sink toward the now leafless tree line; while glittering, the stars of Orion, facing the charge of Taurus, the Bull, rise above the bare branches.

In the south, the large dim groups of the "watery" autumn sky barely attract the casual stargazer. Three oft-heard-of but little-known zodiacal constellations—Capricornus, the Sea Goat, Aquarius, the Water Man, and Pisces, the Fishes—are visible.

Directly overhead, though, are the players in the most famous of all the star legends, that of Cassiopeia, the vain queen, Cepheus, the foolish king, their innocent daughter, Andromeda, chained to a rock by the sea, and Perseus, the hero, her rescuer. There, too, are two creatures who figure in this drama—Cetus, the sea monster, and Pegasus, the flying horse, mount of Perseus.

How did these characters find themselves enstarred in the firmament? The story begins when Queen Cassiopeia, represented as a W-shaped star group straight overhead, boasted that she was more beautiful than the Nereids, mermaids, and the attendants of the sea goddess Thetis. They complained of this to Poseidon, god of the seas, who sent great storms and a savage sea monster to ravage the coast of Cepheus's kingdom.

Cepheus consulted the oracle and learned that he could appease the god by sacrificing his daughter Andromeda to the sea monster. She was chained naked to a rock awaiting her doom when the hero, Perseus, flying home on the back of Pegasus after slaying the Gorgon, saw and fell instantly in love with her. He landed and, after consultation with Cepheus and Cassiopeia, agreed to rescue Andromeda if she would be his bride.

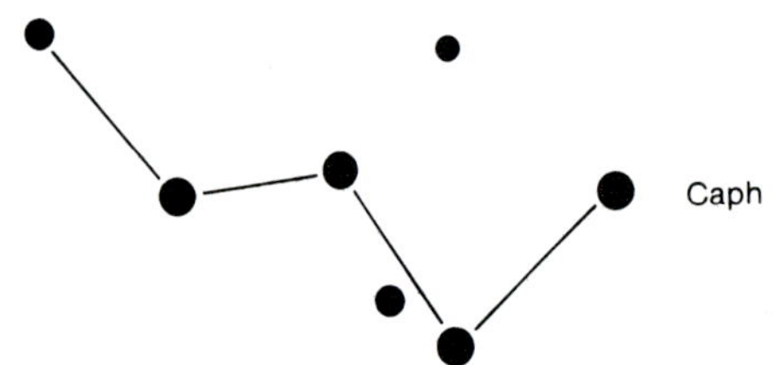

Cassiopeia

On his winged mount, Perseus descended from the sky and cut off the head of the monster Cetus. He then sacrificed a calf, a cow, and a bull to propitiate the gods. Although Andromeda was happy to marry Perseus, indeed, insisted on it, the king and queen broke faith by summoning a former suitor of Andromeda's to the wedding to attack Perseus. To defend himself, Perseus was forced to unwrap the Gorgon's head, which turned the guests to stone. Poseidon set Cassiopeia and Cepheus in the sky, where, for their treachery, they stand foolishly on their heads half of each year.

Cassiopeia is a W- or an M-shaped group, depending on her position in the sky, the image of the queen sitting on her throne. Cepheus, the king, is easiest to see as a stick-figure house much like children draw. Andromeda consists of two diverging lines of stars, her body and legs, crossed by a line of fainter stars, her outstretched, chained arms. She is on the meridian this month. Two distinctive curved lines of stars form Perseus, to the east of Andromeda.

Four bright stars called the Great Square make up the body and wings of the Flying Horse. The head and legs stretch from the Great Square. The classic figure of Cetus, the Sea Monster, consists of an ugly head of five stars rising from a squat body. We show Cetus as a friendly whale, with the five-star figure his tail.

EAST

Even as the autumn drama reaches its fullest overhead, the stars of winter are rising. Mighty Orion, the Hunter, and Gemini, the Twins, now stand above the eastern treetops, while Auriga, the Charioteer, and Taurus, the Bull, are well up. Don't miss the glittering Pleiades or the stately Hyades around Aldebaran, the eye of the Bull.

This map shows the sky as it appears on:

August 23 at 4 A.M.; September 7 at 3 A.M.; September 22 at 2 A.M.; October 7 at 1 A.M.; October 23 at midnight; November 7 at 10 P.M.; November 23 at 9 P.M.; December 7 at 8 P.M.; December 23 at 7 P.M.

NORTH
NORTHEAST
NORTHWEST
URSA MAJOR
URSA MINOR
DRACO
NCP
Polaris
CEPHEUS
Vega
LYRA
Pollux
Castor
CAMELOPARDALIS
AURIGA
CASSIOPEIA
Capella
Deneb
PERSEUS
Caph
CYGNUS
Algenib
SAGITTA
ECLIPTIC
GEMINI
Algol
TAURUS
ANDROMEDA
TRIANGULUM
DELPHINUS
Altair
WEST
Betelgeuse
Aldebaran
The Pleiades
Hamal
PEGASUS
ARIES
Enif
EQUULEUS
AQUILA
PISCES
ORION
Alrischa
Rigel
Mira
ECLIPTIC
LEPUS
CETUS
ERIDANUS
CAPRICORNUS
AQUARIUS
FORNAX
Fomalhaut
PISCIS AUSTRINUS
SOUTHEAST
SOUTHWEST
PHOENIX
SOUTH

THE DECEMBER SKY

With the arrival of cold weather come the bright stars of winter. The southeastern sky now fairly blazes with brilliant stars, while the southwestern sky is dark with the vast, dim constellations of autumn. The very last of summer's sky figures, the Northern Cross in the constellation Cygnus, the Swan, stands upright in the northwestern sky. It is fitting that the Northern Cross is placed this way at this time of year.

On the meridian is one of the smallest of the zodiacal constellations, Aries, the Ram. Consisting of three stars, the tiny grouping nonetheless conveys the essence of a ram, his head down, ready to butt. Centuries ago, the sun crossed the celestial equator in Aries as it moved north in the spring. This place was called the first point of Aries, the vernal (spring) equinox. The slow wandering of the earth's axis has moved the vernal equinox to Pisces, but Aries remains the first constellation of the zodiac in the rhyme:

The Ram, the Bull, the Heavenly Twins,
And next the Crab, the Lion shines,
The Virgin, and the Scales.
The Scorpion, Archer, and the Goat,
The Man who pours Water out,
And Fish with glittering tails.

South of Aries lurks Cetus, the Whale, one of the "heavies" in the Cassiopeia-Perseus legend. Classic renditions show Cetus as a sea monster rearing his ugly head—a circle of five stars below Aries—from the ocean. Our version shows Cetus as a whale, the circle of stars his raised tail, swimming peacefully between Aquarius, the Waterman, and Eridanus, the River.

One peculiar star in Cetus demands special mention: Mira Ceti. This star was among the first known variable stars. It cycles in somewhat over a year from a readily visible second-magnitude star to total naked-eye invisibility and back again. Known as Mira ("wonderful" in Latin), this star is the brightest long-period variable star, visible to the naked eye through at least part of its variation.

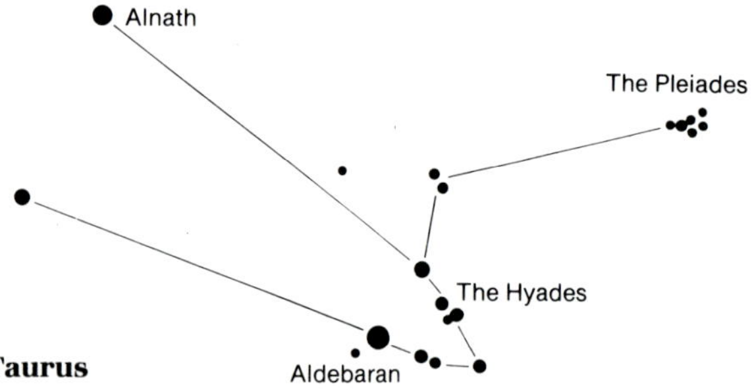

To the south of Cetus and the east of Orion, the Hunter, the River Eridanus flows down the heavens. Fed by a spring at the knees of Orion, the River twists first east, then south, east again, south, west, south, finally dipping below the horizon for most North American stargazers. Eridanus finally ends at first-magnitude Achernar, a star that lies 33° from the south celestial pole.

In the west, Pegasus, the Flying Horse, plunges headlong toward the horizon. Below him, the Water Jar of Aquarius is also sinking. Pisces, the Fishes, twine between the wing of the horse and the back of Cetus, the Whale. Overhead, the figures of Cassiopeia, the vain queen, the foolish king Cepheus, the tender maiden Andromeda chained to the rock awaiting her fate, and her rescuer Perseus remain locked in their timeless drama.

In the northern sky, Ursa Minor hangs by his elongated tail from the polestar. Draco is at its lowest, but the Larger Bear, Ursa Major, once again pokes his nose out of his den and rises in the sky.

But the real action lies to the east where the brilliant stars of winter are now above the horizon. Rigel, Aldebaran, Capella, Castor, Pollux, Procyon, and Sirius encompass all the bright stars in the sky. Three in a row mark the belt of mighty Orion, the Hunter. His faithful dogs, Canis Major and Canis Minor, follow at his heels. As Taurus charges Orion, the Twins stand watching. All the while, Auriga wheels his chariot, heedlessly, across the sky.

This map shows the sky as it appears on:

September 7 at 5 A.M.; September 22 at 4 A.M.; October 7 at 3 A.M.; October 23 at 2 A.M.; November 7 at midnight; November 23 at 11 P.M.; December 7 at 10 P.M.; December 23 at 9 P.M.; January 7 at 8 P.M.; January 23 at 7 P.M.

NORTH
NORTHEAST
NORTHWEST
WEST
SOUTHWEST
SOUTH
DRACO
URSA MINOR
NCP
Polaris
Dubhe
URSA MAJOR
CEPHEUS
Deneb
CYGNUS
CASSIOPEIA
CAMELOPARDALIS
Caph
CANCER
Castor
Pollux
PERSEUS
AURIGA
Capella
Algenib
ANDROMEDA
ECLIPTIC
Algol
PEGASUS
Enif
TAURUS
TRIANGULUM
Pro-cyon
CANIS MINOR
GEMINI
The Pleiades
Hamal
ARIES
PISCES
Betelgeuse
Aldebaran
ORION
ECLIPTIC
Alrischa
AQUARIUS
ONOCEROS
Mira
Sirius
Rigel
CETUS
LEPUS
ERIDANUS
CANIS MAJOR
FORNAX
COLUMBA

2. Choosing and Using a Telescope

All of the stars on the monthly sky maps in chapter 1 and many of the stars and deep-sky objects shown on the star charts in chapter 3 are visible to the naked eye. But to appreciate the details of their appearance you'll need something to aid your vision, either a binocular or a telescope. If for now you're using the charts for naked-eye viewing, you can skip this chapter and go straight to chapter 3. But if you own a binocular or telescope, or if you're planning to buy one, read this chapter first.

There are three main types of telescope: refractor, reflector, and catadioptric. *Refractors* use a lens to form an image of the sky. *Reflectors* have a precision mirror that forms an image of the sky. *Catadioptrics* use both lenses and mirrors to form their image.

In most telescopes, objects appear upside down.

With all three types, you view the image with a lens called an *eyepiece*. The eyepiece determines the power, or magnification, of a telescope. Eyepieces are interchangeable. You can switch from 30× magnification to 600× simply by removing one eyepiece from the telescope and putting on another.

However, it's the diameter of the main lens or mirror, or the telescope's *aperture*, that determines the highest magnification that will be useful. This is usually about 50× for each inch of aperture. The aperture also determines how much light the telescope can collect, how bright the view through the telescope will look, and how faint a star you will be able to see with it.

The larger the aperture, the more a telescope can do. A telescope with an aperture between 2.4 inches (60 mm) and 5 inches (125 mm) is just fine for a beginner. A telescope having an aperture between 6 inches (150 mm) and 11 inches (280 mm) can do enough to keep an ardent amateur astronomer happy for life. Larger telescopes tend to be rather specialized.

In the following pages, I'll discuss the advantages and drawbacks of different kinds of telescopes and mountings in more detail.

BINOCULARS

A binocular is two small refractor telescopes side by side, one telescope for each eye. Throughout this book, I've stressed how many sky objects you can see with binoculars. Binoculars are easy to use, versatile, and portable. The whole family can enjoy them for sports, hunting, or bird watching, as well as for astronomy. A good binocular makes an excellent introductory telescope for the whole field of astronomy—though you will probably eventually want to buy a telescope.

REFRACTORS

Refractors make excellent first telescopes. The classic 60-mm (2.4-inch) refractor is large enough to show lunar craters, Jupiter's satellites and cloud bands, the rings of Saturn, quite a few double stars, and bright deep-sky objects. On the whole, though, a refractor of 80 mm (3¼ inches) or 100 mm (4 inches) aperture would be a better choice. These give you the power to reveal the polar caps of Mars, split lots of double stars, and "discover" hundreds of star clusters and nebulae.

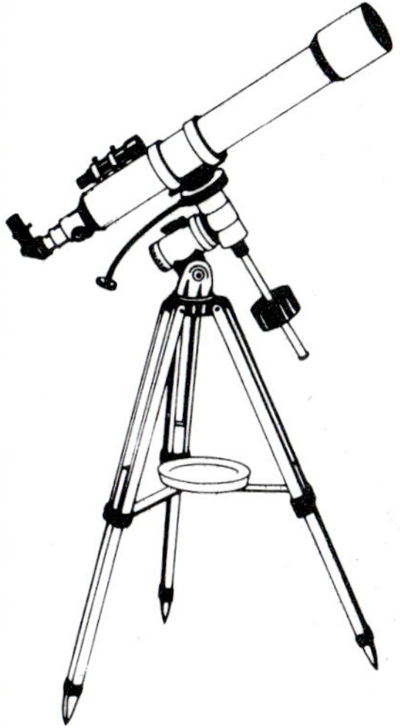

Refractor telescope

There are three types of refractors. *Achromatic* refractors (achromats) have lenses made from two different types of glass. *Apochromatic* refractors (apochromats) have three different types of glass. *Fluorite* refractors employ a special optical material called fluorite in their lenses.

Achromatic refractors use two types of glass because a lens made with only one type of glass forms an imperfect image. Correcting this imperfection (called chromatic aberration) allows an achromatic lens to form a nearly perfect image.

Apochromatic refractors are even better "color-corrected" than achromatic lenses. Although apochromatic refractors are more expensive than achromatic refractors of the same aperture, the added cost is worth it when the aperture is 100 mm (4 inches) or larger.

The unusual optical properties of fluorite permit near-perfect color correction in fluorite refractors, but they are very expensive compared with achromatic and apochromatic refractors.

REFLECTORS

Reflectors are larger and cost less than refractors. Mirrors can form images just as well as lenses do, and unlike lenses they don't suffer from chromatic aberration. Reflectors from 6-inch (150-mm) to 10-inch (250-mm) aperture can provide years of satisfaction for a reasonably low price. In general, reflectors need a little more know-how to use, and require more maintenance than other types of telescopes.

Most reflectors use the Newtonian optical system, invented by the English scientist Sir Isaac Newton and named after him. A large concave mirror at the bottom end of the telescope reflects light to a small flat mirror near the front of the tube, then out a hole at the side. You look through an eyepiece located near the top of the tube.

Equatorial Newtonian reflector

Newtonian reflectors are fine all-purpose telescopes. You can enjoy superb views of the moon and planets with a Newtonian reflector, then put in a different eyepiece and go looking for star clusters or nebulae.

Dobsonian reflectors are Newtonian reflectors on low-cost mountings (see "Mountings" below). Dobsonians often have apertures up to 17 inches (440 mm) or sometimes even more. Their enormous light-gathering power makes them perfect for observing star clusters, nebulae, and galaxies.

CATADIOPTRIC TELESCOPES

Catadioptric telescopes are the most popular telescopes made today. In astronomy, you are most likely to encounter the Schmidt-Cassegrain catadioptric telescope, or SCT. Although SCTs are more expensive than reflectors of the same aperture, they are considerably more portable. A typical 8-inch aperture (200-mm) SCT has a tube only 18 inches long and weighs under 20 pounds. SCTs offer the beginner good views of the moon and planets, double stars, and deep-sky objects.

The SCT optical system consists of a thin corrector lens, a primary mirror, and a secondary mirror mounted on the inside of the corrector plate. Because the optical path is folded inside the telescope, the tube can be remarkably short.

Catadioptric (Schmidt-Cassegrain) telescope

MOUNTINGS

It is as important to get a good mount as it is to find the right telescope. A telescope mounting should support the tube solidly, without allowing the tiniest bit of wobble, and let you point it at any part of the sky. Look for a mounting that is ruggedly made and easy to use.

In modern commercial telescopes, you'll usually find one of four types of mountings: camera tripods, altazimuth mounts, equatorial mounts, and Dobsonian mounts.

Camera tripods are sometimes sold for use with telescopes. They aren't too bad for daytime use with "bird-watcher's" telescopes. Avoid them for astronomy, except to support binoculars. For telescopes, they are much too shaky.

Altazimuth mounts move the telescope on vertical and horizontal (up-and-down and back-and-forth) axes, but the stars move at all angles across the sky. This means that you will have to turn two knobs to follow a star as it moves across the sky. With small telescopes and low-to-medium magnification, this is no problem.

Equatorial mounts have a polar axis that is parallel to the earth's axis of rotation, and another perpendicular to it. With a drive motor, the mount will automatically turn your telescope to follow the stars. This is very convenient, especially at high magnification.

Equatorial mountings often have "setting circles" using equatorial coordinates for pointing the telescope at celestial objects. Called *right ascension* and *declination*, equatorial coordinates are used for locating stars just as longitude and latitude are used for locating places on our planet Earth. However, you don't need setting circles for finding things in the sky. The charts in this book are sufficient.

Dobsonian mounts are smooth-turning mounts for large reflectors. To find celestial objects with a Dobsonian, you locate the object in the sky and point the telescope at it. To follow a star as it moves across the sky, you gently push on the telescope tube. Dobsonian mounts are entirely satisfactory for deep-sky observing.

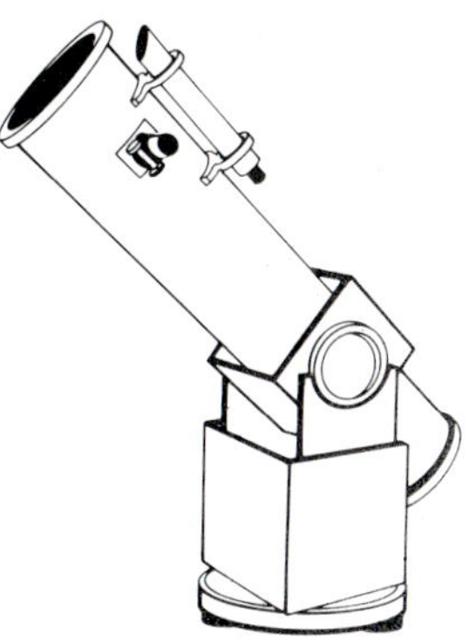

Dobsonian reflector

VERSATILITY, SIZE, AND WEIGHT

The right telescope and mount is a personal choice. However, here are a few hints I hope can help you decide what telescope is best for you.

In a nutshell, refractors, especially the long-focus achromats and ultraperfect apochromats, give the most satisfying views of the moon and planets. Medium- and large-sized Newtonians provide the maximum performance per dollar and also rank very high in versatility, but they're bulky. Large Dobsonian telescopes are tops for observing deep-sky objects. Schmidt-Cassegrain telescopes offer a bit of everything, and they're both compact and portable.

Avoid camera tripods; they aren't for stargazers' telescopes. A well-made altazimuth mount is quite adequate for a small refractor. For a refractor with a 3-inch (80-mm) or larger aperture, an equatorial mount is well worth the extra expense. For a Newtonian reflector, you should have an equatorial mount with a drive motor unless your primary interest is deep-sky observing. In that case, a Newtonian with a Dobsonian mount will serve your needs well. Many Schmidt-Cassegrain telescopes include an equatorial mounting with a built-in drive motor.

Remember that the large-aperture telescope that can show you everything you want to observe may not actually be the best for you. A telescope has to be carried out and set up each time you use it. If it's too big and bulky for you, you won't enjoy using it. Furthermore, if you live in the city and have a yen for deep-sky observing, you will have to drive quite some distance to find dark skies. Do not forget that the telescope's size and portability are important factors. The physical height, length, and weight of the instrument must be part of your thinking. Read the manufacturer's specifications carefully before getting a big telescope.

One final thought: As long as you select a well-made telescope from a reputable maker, you really can't go far wrong.

EYEPIECES AND ACCESSORIES

As mentioned earlier, eyepieces determine a telescope's magnification. You should have one eyepiece for low magnification, onc for medium, and one for high power. *Low* means 5× to 7× per inch of aperture, or about 36× for a 6-inch telescope and 48× for an 8-inch telescope. *Medium* magnification means 16× per inch of aperture, and *high* magnification about 40× per inch of aperture. Magnification greater than 50× per inch of aperture is useless, or "empty," magnification.

You can find the magnification any eyepiece will give by dividing the focal length of the telescope (which you can find in the manual or engraved on the telescope tube) by the focal length of the eyepiece (which you will find engraved on the eyepiece). For example, if the focal length of the telescope is 2,000 mm and the focal length of the eyepiece is 32 mm, then the magnification of the telescope with that eyepiece is 62×.

If you have a refractor or a Schmidt-Cassegrain telescope, be sure you get a *star diagonal*. A star diagonal attaches between the telescope and eyepiece, and changes the viewing angle by 90°. It allows you to look through the telescope when it is pointing upward without getting a stiff neck.

A finder telescope is a necessity because the sky area you see with your main telescope is quite small. A finder is a low-power telescope mounted piggyback on the main telescope, to make it easier to get celestial objects into the field of view. Most telescopes include finders, but if yours doesn't, then order one and install it.

USING YOUR TELESCOPE

Set up your telescope in a place as far as you can from artificial lights, and where you are sheltered from the wind. Street and house lights prevent your eyes from becoming sensitive to the faint light of celestial objects, and wind may be chilling and can shake your telescope. Avoid concrete or asphalt surfaces as these store heat from the day and release it at night. The rising warm air destroys the fine definition a good telescope can provide.

Set up your telescope according to the directions that come with it. Telescopes on altazimuth and Dobsonian mountings require no special positioning, but equatorials must be aligned so that the polar axis points to the north celestial pole, near the star Polaris. Beginners often find equatorials confusing to set up, so follow the manufacturer's directions carefully. If your telescope has a drive motor, you will have to run a power cord out to it.

Have your astronomer's flashlight handy (see chapter 1). With it you will be able to read the sky maps and star charts in this book without dazzling your eyes. It takes 15 to 45 minutes for your eyes to become fully adapted to the dark. Dim orange or red light does not hurt your night vision nearly so much as white light does.

A small table and a chair are useful. Keep your flashlight, book, any spare eyepieces, and a notebook for recording your observations on the table. A chair lets you sit to observe whenever the position of the telescope and eyepiece allow it.

You are now ready to observe. Place a low-power eyepiece in the eyepiece holder and point the telescope at a bright star. Even with a low-power eyepiece, the field of view is small and you may need to search a bit before you find the star. With practice you will become adept at finding celestial objects.

Once you have the star, carefully center it. Look through your telescope's finder, and if the star is not centered on the finder's cross hairs, turn the adjustment screws that hold the finder until it is aligned with the main telescope. Locating faint stars and deep-sky objects will be much easier if you use a finder.

Turn the focus knob until the star is a tiny, bright point of light. You can observe with either eye, as you prefer. It may help reduce eyestrain if you can keep the other eye open, but this isn't necessary. If you wear glasses, you may have to remove them to look into the tiny openings of high-power eyepieces. If you observe with contact lenses, be careful they don't pop out.

If your telescope does not have a drive motor, the star you have found will drift out of the field in a minute or two, and you will need to bring it back. Because telescopes invert the image of the sky, the star will "move the wrong way" as you try to recenter it. You will soon learn how to keep a star in the field without thinking about it, especially if your telescope has a Dobsonian mounting. A motor-driven equatorial automatically follows the stars provided its polar axis is correctly aligned.

When you want higher magnification, first center the star in the low-power eyepiece. Gently remove the low-power eyepiece (without shaking the telescope) and slip in an eyepiece that gives the magnification you want. You will probably need to refocus when you change eyepieces.

It is more difficult to find celestial objects at high magnification, more difficult to keep them in view, and easier to bump the telescope and lose the object entirely. Atmospheric turbulence is more annoying at high magnification, too. So always use the lowest magnification that shows you what you want to see.

Jupiter, for example, is best seen between 120× and 180×, and Saturn and Mars seldom show any more with magnifications over 200. You will use powers between 30× and 200× with the moon, depending on what aspect of it you want to observe. For open star clusters and galaxies, a magnification between 30× and 60× is often best; for globular clusters, 150×. For close double stars, use your highest power.

Celestial objects are best when they are highest. Atmospheric turbulence is worst near the horizon, and the Earth's atmosphere absorbs the most light there, too. When you are observing high overhead with a refractor or a Schmidt-Cassegrain telescope, use a star diagonal so you don't have to tilt your head into an uncomfortable position.

Celestial objects are best seen when they're highest in the sky.

To search for faint celestial objects, learn to use averted vision. Instead of staring directly at an object, look to the side of the field while keeping your attention where the object should be. Your eyes can detect fainter light when you aren't looking directly at it, so the object may suddenly pop into view.

Finally, do not be dissatisfied with first impressions. Learning to use a telescope, gaining the necessary observing skills, and coming to appreciate and understand what you are seeking take time. Don't hurry. Give yourself enough time to discover the stars.

3. Exploring the Constellations

The twenty-three star charts that follow in this chapter show you the constellations in more detail than the monthly sky maps in chapter 1. For a particular region of the sky, each chart names key stars and gives you the location of faint stars, star clusters, nebulae, galaxies, and other fascinating deep-sky objects. To aid you in finding your way around the sky and identifying the constellations, there is a wide overlap between adjacent charts. Opposite each chart you'll find a description of stars and deep-sky objects, constellation by constellation, that you can see using binoculars or a telescope.

Like the sky maps, the star charts are arranged by time and season. The charts show the stars highest in the sky, and therefore in the best position for observing. On the page facing each chart is a list of these dates and times.

There are three types of charts: ones that show the sky overhead, ones that show the northern sky, and ones that show the southern sky. The overhead charts show the sky as it looks when you are facing south, looking high above the horizon. The southern sky charts show the sky you'll see when you look south, and they include the southern horizon. The northern charts show the sky as it looks when you face north and look high above the horizon.

If you're using a chart on a date and time given on the list next to it, you need only face the correct direction, hold this book up, and compare the chart directly to the sky. You can use a chart at a time before the hour specified for a given night, but the same stars will be lower and to the east. After the appointed hour, the same stars will be to the west.

These charts are much more detailed than the sky maps in chapter 1. In order to see what's on them clearly, it's important to observe the sky from a good site. City lights brighten the sky, making it hard to see faint stars. Astronomers call this "light pollution" —and wish modern civilization didn't shine so many lights into the

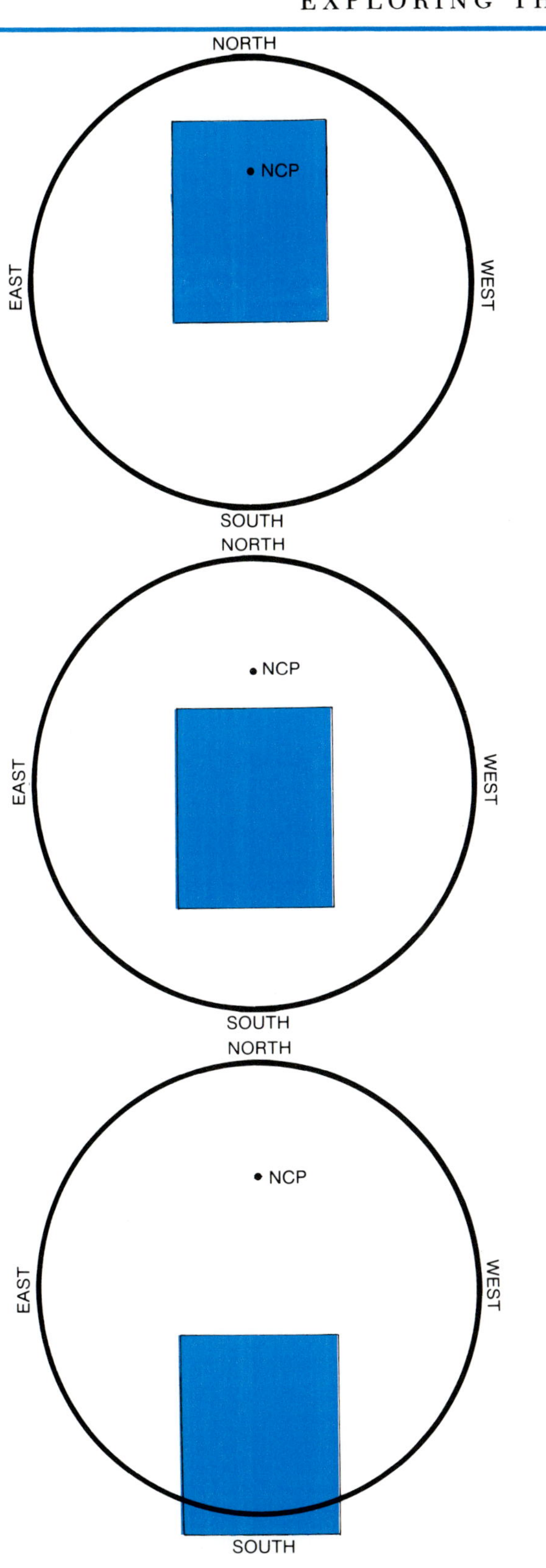

WHAT PARTS OF THE SKY THE CHARTS SHOW

Facing north (star charts 4, 9, 16, and 21)

Overhead, facing south (star charts 1, 2, 5, 7, 10, 11, 13, 17, 19, and 22)

Facing south (star charts 3, 6, 8, 12, 14, 15, 18, 20, and 23)

sky at night. Darkness is important for three reasons: (1) artificial lights are so bright that their glare "blinds" you; (2) lights make the whole sky bright so that faint stars are washed out; (3) bright lights interfere with the dark adaptation of your eyes. To enable you to read the star charts in the dark, be sure you have your "astronomer's flashlight" (described in chapter 1) with you so that you can see the charts without ruining your night vision.

Each chart shows about one-sixth of the entire visible sky. You'll soon develop a feel for how stars on the chart appear in the night sky. For starters, though, think of the real constellations as looking twice as big in the sky as they do on the charts.

NAMES OF STARS AND DEEP-SKY OBJECTS

On these charts you'll find the names of the stars. Many bright stars have proper names, often an anglicized Arabic word for the part of the constellation the star represents. (Because the Arabs preserved Greek learning during Europe's Dark Ages, a lot of Arabic words crept into western astronomy.) For example, the name Alrischa (sometimes also spelled Al Rischa), in Pisces (chart 22), means "the knot" in the cord that joins the two fish. Many of the Arabic names begin with *al*, which means "the."

Most bright stars have Greek letter names. These are called *Bayer letters* after Johann Bayer, who used Greek letters to label the stars in his star atlas, the *Uranometria*, published in 1603. Bayer's Greek letters have been used ever since. Bayer usually assigned the first letter in the Greek alphabet, alpha (α), to the brightest star in a constellation; the second letter, beta (β), to the second brightest; gamma (γ) to the third, and so on. But not always. In Ursa Major, for example, Bayer named the stars in alphabetical order from one end of the constellation to the other! (If you don't know the Greek alphabet, refer to the table on page 45. This table also appears in the appendix on page 119, to make it easy to find when you're outside at night. Greek letters are spelled out in the text, with the symbol given in parentheses.)

A star's full name includes both the Bayer letter and the name of the constellation it's in. The brightest star in Taurus, the Bull, is called alpha (α) Tauri. The constellation name takes the Latin gen-

The Greek Letters

α	alpha	η	eta	ν	nu	τ	tau
β	beta	θ	theta	ξ	xi	υ	upsilon
γ	gamma	ι	iota	ο	omicron	φ	phi
δ	delta	κ	kappa	π	pi	χ	chi
ε	epsilon	λ	lambda	ρ	rho	ψ	psi
ζ	zeta	μ	mu	σ	sigma	ω	omega

itive case ending, which means "of" in English. Thus alpha Tauri means "the alpha star of Taurus."

When two or more stars lie close together, they may have the same Greek letter name with superscripts to distinguish them. In Aquarius (chart 20), for example, we see a little group of stars named psi-1 (ψ^1), psi-2 (ψ^2), and psi-3 (ψ^3). Nearby are omega-1 and omega-2 (ω^1 and ω^2), and tau-1 and tau-2 (τ^1 and τ^2). When there weren't enough Greek letters to go around, Bayer also used lower-case letters from our alphabet; in Puppis (sky chart 6), you'll see a, b, and c Puppis.

When astronomers began discovering that some stars vary in brightness, they labeled these unusual objects using upper-case Roman letters. Since Bayer had used letters up to "q," the first variable in each constellation was named R, the next S, and so on. In Corona Borealis, you'll see stars labeled R and T. After Z, new variables were assigned two-letter designations—RR, RS, RT—until the two-letter combinations were exhausted. Subsequent variables are named V followed by a number.

Stars are numbered, too. When John Flamsteed, the English Astronomer Royal, compiled his definitive star atlas, he marked the stars in it with numbers. This system is confusing because a star may have a proper name, a Bayer letter, and a Flamsteed number all at the same time. The brightest star in Orion, Betelgeuse, is named not only alpha (α) Orionis, but also 58 Orionis. But astronomers use the *Flamsteed numbers* only for stars that don't have Bayer letters. If a star has a proper name or Bayer letter, we do not show its Flamsteed number on the charts.

Deep-sky objects are indicated by a set of symbols that have become standard in modern star atlases. These symbols are shown in the paragraphs that follow, and in a key in the Appendix on page 118.

Deep-sky objects are marked with a *Messier number*, a *New General Catalog* (NGC) number, or an *Index Catalog* (IC) number. Messier numbers date back to Charles Messier, an astronomer who searched for comets. He compiled a list of sky objects to avoid because they might be confused with comets, but people have been looking at them ever since. Messier numbers are written M or M- followed by a number. M31 is the Andromeda galaxy, the closest galaxy to ours. There are 110 Messier objects.

Messier didn't see everything. As astronomers, particularly William and John Herschel, explored the heavens, they found more and more sky objects. In 1888, John Dreyer published a list of nearly 8,000 objects as the *New General Catalog of Nebulae and Clusters of Stars*, abbreviated NGC. In 1895 and 1908, Dreyer added two extensions, called *Index Catalogs*, to the NGC, with over 5,000 more objects. When astronomers talk about something listed in the *New General Catalog* or *Index Catalogs*, they call it NGC or IC plus its number.

ASTRONOMY MEASUREMENTS

Astronomy, like every field, has its own jargon. For example, astronomers use a peculiar system of "magnitudes" to describe how bright celestial objects appear. Most bright stars are called first magnitude. Somewhat dimmer stars are second magnitude, still dimmer ones third magnitude, and so on. Each magnitude is a brightness step of the same size. The faintest star most people can see by eye is sixth magnitude, but a small telescope will show stars as faint as tenth or eleventh magnitude. The funny thing about magnitudes is that they're backwards. A large number for the magnitude means a star is faint, and a small number means a star is bright.

Each magnitude represents a brightness step of 2.5 times. Five magnitudes amounts to a brightness factor of 100 times. Astronomers have extended the brightness scale beyond first magnitude to zero magnitude, –1 magnitude, and so on, and they've added decimals. The full moon is much brighter than a star; it's –12.6 magnitude. The brightest star in the sky is Sirius; its magnitude is –1.5. The charts show every star brighter than 4.5 magnitude plus selected fifth- and sixth-magnitude stars. The scale of magnitudes opposite is repeated on page 119 to make it easy to check in the dark.

Magnitude Scale for Star Charts

6	sixth
5	fifth
4	fourth
3	third
2	second
1	first
0	zero
-1	Sirius

The Brightest Stars

Name	Constellation	Magnitude
Sirius	Canis Major	–1.5
Canopus	Carina	–0.7*
Alpha Centauri	Centaurus	–0.3*
Arcturus	Boötes	–0.0
Vega	Lyra	0.0
Capella	Auriga	+0.1
Rigel	β Orionis	+0.1
Procyon	Canis Minor	+0.4
Achernar	Eridanus	+0.5*
Betelgeuse	Orion	+0.5 (variable)
Hadar	β Centauri	+0.6*
Altair	Aquila	+0.6
Aldebaran	Taurus	+0.8 (variable)
Acrux	Crux	+0.9
Antares	Scorpius	+1.0 (variable)
Spica	Virgo	+1.0
Pollux	β Geminorum	+1.1
Fomalhaut	Piscis Austrinis	+1.2
Deneb	Cygnus	+1.3
Mimosa	β Crucis	+1.3
Regulus	Leo	+1.4

All stars listed are the α stars in their constellations except as marked β.
*Visible only in the Southern Hemisphere

To measure angles in the sky, astronomers use degrees, minutes, and seconds of arc. You're probably familiar with degrees: they're the small divisions on a protractor. There are 360° in a whole circle, and 90° in the angle between the horizon and the zenith, straight overhead. The two end stars in the Big Dipper are 5° apart in the sky. In angular measure, each of the charts is 55° from side to side and 75° from top to bottom.

For angles smaller than 1°, astronomers divide a degree into 60 minutes of arc (arcminutes). The moon occupies an angle of 30 arcminutes, or ½°. For still smaller angles, such as the angle between the components of a double star, we use seconds of arc (arcseconds). An arcsecond is 1/60 of an arcminute, or 1/3600 of a degree. The planet Jupiter occupies an angle of 45 arcseconds. A dime fills an angle of one arcsecond when viewed from one mile away.

Estimating sky angles

Measuring small angles

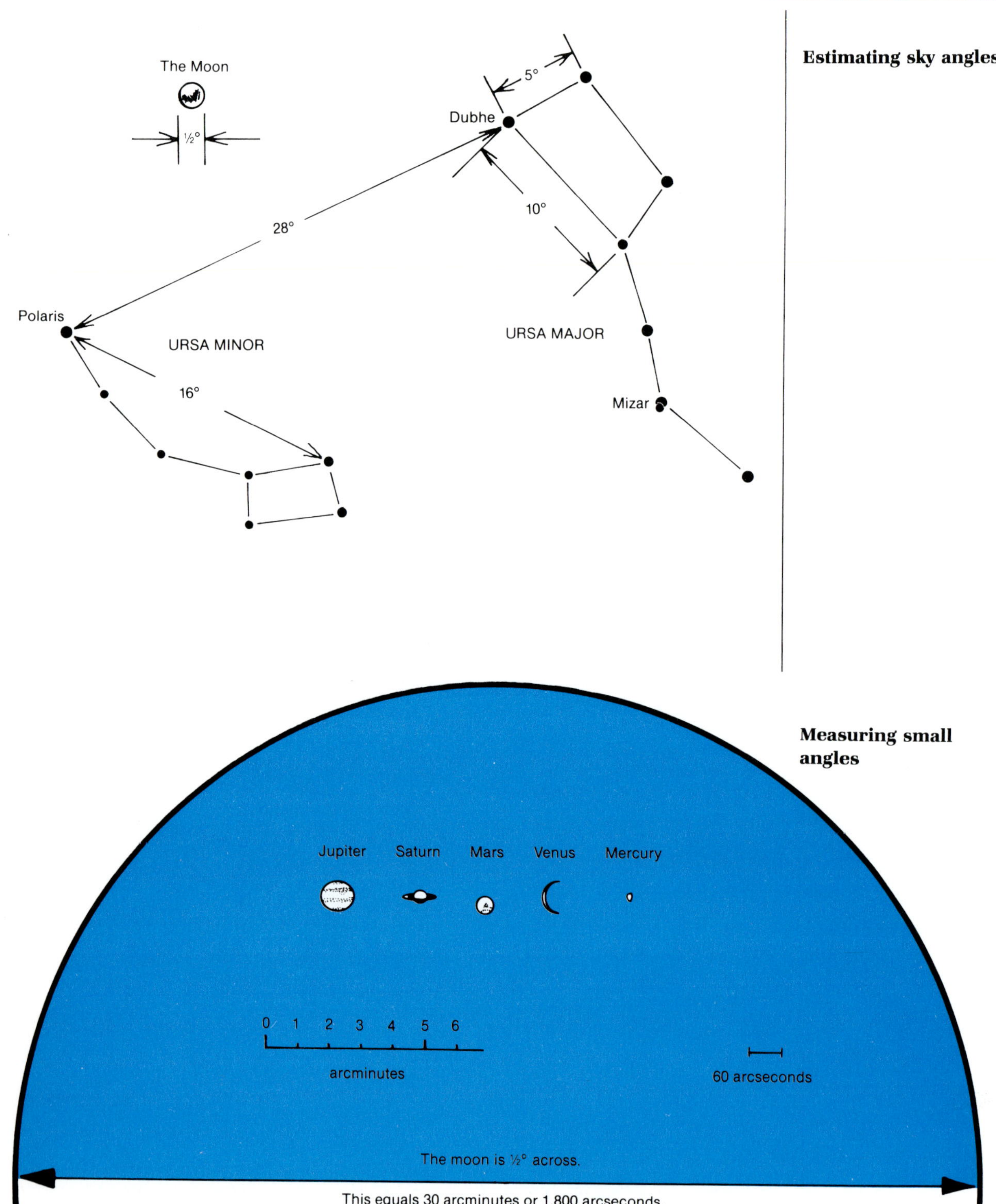

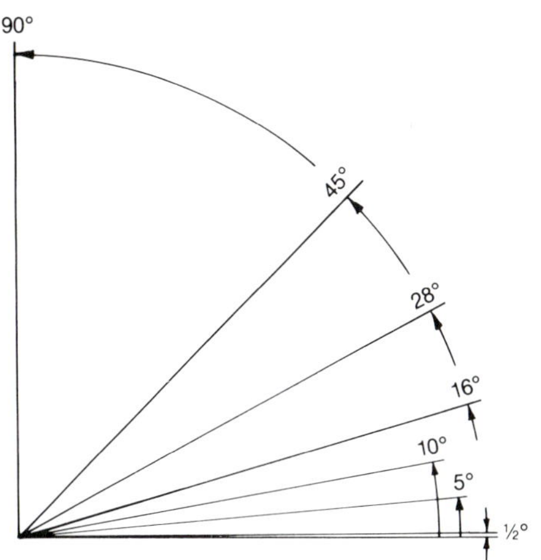

Angles: the astronomer's yardstick. Astronomers describe distances between objects in the sky using angles measured in degrees. There are 360 degrees in a complete circle.

DOUBLE STARS

If you just scan your telescope across the sky, you may notice pairs of stars that seem unusually close. Sometimes this is mere happenstance. Pairs of stars that just happen to be close together in the sky are called *optical doubles*. However, other stars that appear close actually do orbit each other, taking from a few years to a few thousand years to complete a single orbit. The symbol for a double star is a dot with a line through it, like this: -•- in white.

The first double star discovered is also one of the prettiest. It's called Mizar, or zeta (ζ) Ursae Majoris, and it's the middle star in the handle of the Big Dipper (see chart 9). If you have good eyes, you'll see a faint star called Alcor right next to Mizar. Mizar and Alcor form an optical double. When you examine Mizar itself with a telescope, you'll see that what looked like one star really consists of two very close stars. Those two stars orbit each other. Mizar is a physical double, or *binary* star.

A particularly beautiful, bright, and easy-to-find double is Albireo,

Mizar: a double star

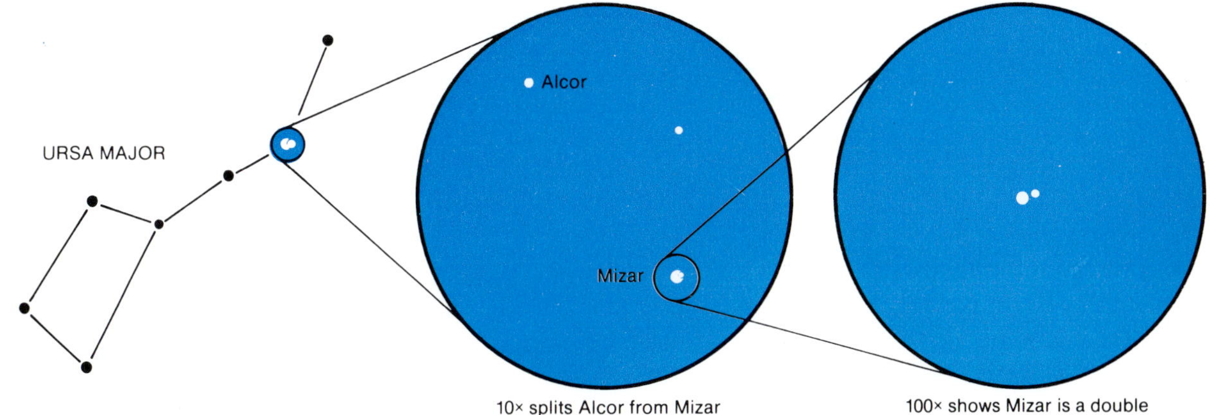

10× splits Alcor from Mizar

100× shows Mizar is a double

the end star in Cygnus. It's marked as beta (β) Cygni on chart 17. Albireo's two stars are farther apart than Mizar's are, but the striking thing about them is the color contrast between them. The brighter star is yellowish or orange, while the fainter star is blue.

VARIABLE STARS

Variable stars are stars that change in brightness. Some change only a small amount in brightness, but others change a great deal. Some change rapidly and others vary slowly.

Long-period variables, or Mira variables, named after the star omicron (o) Ceti, also called Mira Ceti (see chart 23), vary slowly and over a wide range of brightness. Mira variables are old, cool stars that slowly pulsate as a result of changes deep inside their cores.

Cepheid variables are those that vary like the star delta (δ) Cephei (see chart 16). They regularly brighten and dim a magnitude or so in a few days or weeks. Cepheids are extremely bright stars, and their light output and period of variability are related. As a result, astronomers use Cepheids to find the distances to other galaxies. Polaris is a Cepheid.

Binary stars may eclipse as they orbit one another, and when they do, we see their light vary. Algol, or beta (β) Persei (chart 22) is such a variable. Eclipsing binaries are usually very regular in their variation.

R Coronae Borealis (see chart 11) is another prototype variable. This enormous ruddy star is so cool that carbon in its outer atmosphere solidifies and blocks the light from deeper down. R Cor Bor then abruptly fades for several weeks.

T Coronae Borealis (also chart 11) is a recurrent *nova,* a star that suddenly brightens hundreds or thousands of times. There are several types of novae. Some flare up regularly and others do not, but all blaze out when an excess of hydrogen "fuel" suddenly ignites on an old star.

Finally, there are *supernovae,* stars that blow themselves apart. In 1572, Tycho's supernova shone in Cassiopeia (chart 21) as bright as Venus, at magnitude –4! The nebula M1 in Taurus (chart 1) is the remnant of a supernova that blew up in A.D. 1054, and the Veil nebula in Cygnus (chart 17) is a remnant of a supernova that happened at least 10,000 years ago.

On the charts, variable stars are shown by an open circle ⭘ in white, or if the range of variation is too small to be visible, as an ordinary white star dot.

STAR CLUSTERS

Scattered along the Milky Way are dozens of loose aggregations of stars called *star clusters*. There are two kinds of clusters, and they don't look a bit alike through a telescope.

The easiest clusters to recognize are *open clusters*, which include a few dozen to a few hundred stars that seem to be very loosely stuck together. With a small telescope or binocular, an "average" star cluster looks somewhat like a small scattered pile of salt on a sheet of black paper. Most lie fairly close to the plane of our galaxy, so they appear in the Milky Way. The symbol for an open star cluster is ⁘.

By far the most stunning open cluster in the sky is the Pleiades, a cluster in the winter sky (see chart 1). Once you've located the Pleiades by eye (it's fairly easy to see even from city areas), point your binocular or telescope at it. You'll be amazed: there'll be seven bright bluish stars glowing like jewels set in a field of dozens of fainter stars and hundreds of tiny ones filling the field.

The other kind of star cluster is the *globular cluster*, a far larger and more distant ball of hundreds of thousands of stars. Because they are so distant and the stars in them are so faint, small telescopes don't show the individual stars, but show only a gentle glow. The symbol for a globular star cluster is ⊕.

The best and brightest northern globular lies in the constellation

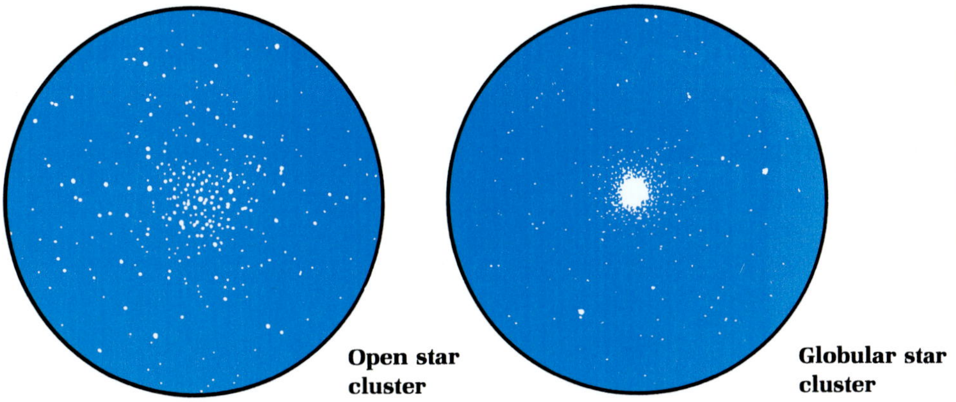

Open star cluster

Globular star cluster

Hercules, shown on chart 13. You'll see this globular labeled M13, the thirteenth object on the list compiled by the famous observer Charles Messier. At low power, it looks like a soft fuzzy blob. If you happen to be under a good, dark sky, try a higher power. Look beside the cluster rather than directly at it to throw the image of the cluster in the most sensitive part of the retina, and you'll see that M13 is made up of thousands of just barely perceptible separate stars.

NEBULAE

Some of the clusters you'll find if you go poking along the rich fields in the Milky Way seem to have a fuzzy haze around them, like a little cloud. Early sky explorers noticed this and called them "nebulae," Greek for clouds.

Emission nebulae are now known to be vast hydrogen clouds from which the cluster stars form; the gas is lit by the light from the stars. Emission nebulae are marked on the charts with the symbol □, except for a few large nebulae that are drawn to size. The best known and most impressive emission nebula lies in the sword of the bright constellation Orion (see chart 2). Point your telescope at the stars of Orion's sword—and right away you'll see an unusual glow between them. That's M42, the Orion nebula. It will have a few bright stars at its center. The nebula sweeps around the central stars in a dramatic swirling fashion! You might not find it very impressive if you're in the city, but from dark country skies, the nebula sprawls across the field of view.

Second only to the Orion nebula is a beautiful summertime nebula called the Lagoon nebula (M8). The Lagoon lies in the Milky Way in Sagittarius (chart 15). Find the Sagittarius "teapot" star pattern, then point your telescope at the "spout" and raise it a few degrees. The Lagoon will almost surely be shining softly in the eyepiece, wrapped around a faint star cluster.

Another type of nebula is the *planetary nebula*. Through a telescope a planetary looks a bit like a very dim planet. Planetaries are actually expanding globes of gassy debris left over when a star explodes. Planetary nebulae are marked on the charts by the symbol ◎.

Two nebulae in particular are worth seeing: the famous Ring nebula (M57), which looks like a smoke ring, and the Dumbbell nebula (M27). The Ring is in the bright constellation Lyra, between the

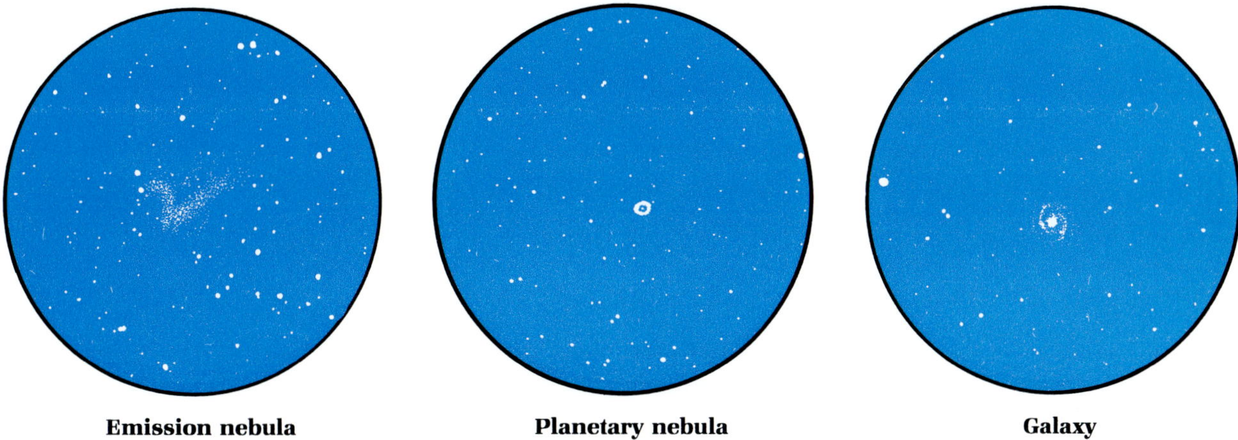

Emission nebula **Planetary nebula** **Galaxy**

two southern stars in the lyre. The ring is a tiny oval glow at low power. Since the Ring nebula is fairly bright, try a higher magnification: You'll see it become a small but very distinct smoke ring (or some say doughnut) shape. The Dumbbell is hard to find from a city because you need to see the dim constellation Delphinus to find it easily (see chart 17). Scan between the star Albireo and the "diamond of Delphinus." You'll find the Dumbbell about one-third of the way between them, and closer to the Albireo end. It's an excellent opportunity to practice with a star chart, and the reward of finding this nebula makes the effort worth it!

Supernova remnants are still another type of nebula. They are the debris left over when a hot, young star quite literally blows itself to bits. The Crab nebula in Taurus, M1, is a supernova remnant. It is a rather faint glow near zeta (ζ) Tauri. This supernova remnant is marked with the symbol +.

THE MILKY WAY

Stretching around the sky in a complete circle is the Milky Way, a band of light composed of millions of stars in our galaxy. You'll see this band of light arching across the heavens late on summer nights, providing you can get far away from bright city lights.

Besides the countless gemlike individual stars, the Milky Way contains star clusters, nebulae, and clouds of faint stars. The very best fields lie in Sagittarius and Cygnus, and they're quite different. Sagittarius abounds in nebulae and clusters. In ten minutes you can find a dozen (and probably more) Messier-numbered objects! In Cygnus, the sheer number and brilliance of the stars will overwhelm you; there will be thousands in every field of view. The experience is indescribable. You have to discover it yourself—perhaps from the shore of a quiet lake in Canada after the campfire dies down, or from the High Sierras far from human presence.

The Milky Way appears on the star charts as a light patch running through the sky.

GALAXIES

Galaxies are one class of deep-sky objects that every astronomy buff wants to see. However, beyond the brightest few, it takes a large telescope to truly appreciate galaxies. No telescope can show them as photographs do. Most of them look like little fuzzy spots. But they're somehow awesome.

The number one galaxy in the sky is the Andromeda galaxy, 31 on the Messier list. You can see it by eye if you know just where to look in the autumn constellation Andromeda. With a telescope, use the lowest power. M31 is a soft, oval glow that's brightest in the center. It will span the entire field. Near Andromeda are two fainter galaxies. The round one close to the main galaxy is known as M32; the larger but fainter oval one is called NGC 205. How's that for your first galaxy?—three-in-one!

In the spring sky, look for M81 and M82 in Ursa Major (see chart 9). With a low-power eyepiece, these galaxies will be small, somewhat oval glows about one moon diameter apart. M81 is a normal spiral galaxy, but M82 is a very strange exploding galaxy. Even with a small telescope, you'll be able to see that they are quite different.

The charts show nearly one hundred galaxies, all bright enough to find with small telescopes. They are marked with the symbol ○. The thrill of finding galaxies lies not in what you actually see (which isn't much even with the very largest telescopes), but in finding these remote, mind-boggling spirals and ovals and viewing them with your own eyes.

USING THE STAR CHARTS

If you are interested in finding a particular constellation on the charts in this chapter, just look it up in the List of Constellations at the end of the book. The list will direct you to the charts that include the constellation you're looking for.

If you're out looking at the stars and just want to find the chart

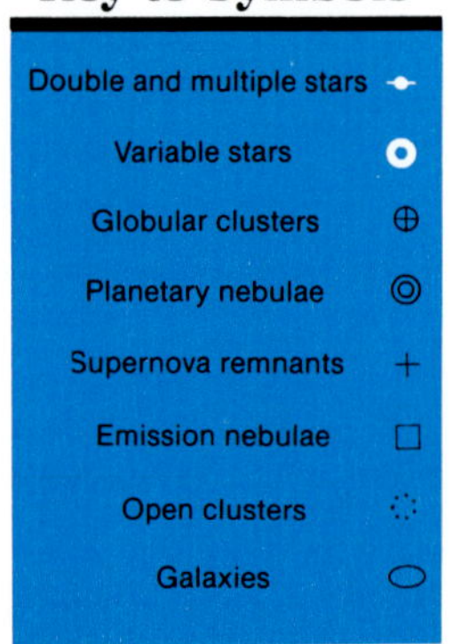

that highlights the stars that are highest overhead at a given time, use the star chart timetable below. This will direct you to the charts that show the most visible stars at any time on a given night. This table is repeated in the Appendix on page 119, where it will be easy to find at night.

The key to the symbols used in the star charts is also repeated in the Appendix.

Which Star Chart to Use

	6 P.M.	7 P.M.	8 P.M.	9 P.M.	10 P.M.	11 P.M.	MIDNIGHT	1 A.M.	2 A.M.	3 A.M.	4 A.M.	5 A.M.	6 A.M.
Jan. 7	21	22	23		1	2,3	4		5,6		7,8		9
Jan. 23	22	23		1	2,3	4		5,6		7,8		9	10
Feb. 7	23		1	2,3	4		5,6		7,8		9	10,11	
Feb. 21		1	2,3	4		5,6		7,8		9	10,11	12	
Mar. 7	1	2,3	4		5,6		7,8		9	10,11	12		
Mar. 23	3	4		5,6		7,8		9	10,11	12		13,14	15
Apr. 7	3	4		5,6		7,8		9	10,11	12		13,14	15
Apr. 22	4		5,6		7,8		9	10,11	12		13,14	15	16
May 7		5,6		7,8		9	10,11	12		13,14	15	16	17
May 23	6		7,8		9	10,11	12		13,14	15	16	17,18	
Jun. 7		7,8		9	10,11	12		13,14	15	16	17,18		
Jun. 22	7,8		9	10,11	12		13,14	15	16	17,18			19
Jul. 7		9	10,11	12		13,14	15	16	17,18			19,20	
Jul. 23	9	10,11	12		13,14	15	16	17,18			19,20		21
Aug. 7	11	12		13,14	15	16	17,18			19,20		21	22
Aug. 23	12		13,14	15	16	17,18			19,20		21	22	23
Sep. 7		13,14	15	16	17,18			19,20		21	22	23	
Sep. 22	13,14	15	16	17,18			19,20		21	22	23		1
Oct. 7	15	16	17,18			19,20		21	22	23		1	2,3
Oct. 23	16	17,18			19,20		21	22	23		1	2,3	4
Nov. 7			19,20		21	22	23		1	2,3	4		5,6
Nov. 22		19,20		21	22	23		1	2,3	4		5,6	
Dec. 7	19,20		21	22	23		1	2,3	4		5,6		7
Dec. 23		21	22	23		1	2,3	4		5,6		7,8	

This table shows you which star charts to use at any time of the night throughout the year. For instance, if you're out on or near August 7 at 9:00 P.M., the stars shown on star charts 13 and 14 will be in view. If you stay out that night until 1:00 A.M., the stars shown on charts 15, 16, 17, and 18 will come into view high overhead.

On this table, and on all the smaller tables facing the star charts, times for January, February, March, November, and December are given in standard time. Times for April, May, June, August, September, and October are given in daylight savings time.

STAR CHART 1
Overhead, facing south

Centered on one of the starriest regions in the sky, this chart shows four brilliant winter constellations and a swath of the Milky Way that abounds in star clusters. On this page we explore Taurus, Gemini, and Auriga. The fourth bright constellation, Orion, is also filled with interesting celestial sights; it is covered in star chart 2.

Taurus, the Bull

Taurus is one of the brightest of the zodiacal constellations. It boasts two naked-eye star clusters—the Hyades and the Pleiades—as well as the remnant of a supernova that in A.D. 1054 shone as bright as the planet Venus. Today this remarkable object is a ninth-magnitude nebula known as the Crab nebula, M1. It lies 1° north of the star zeta (ζ) Tauri, a faint smudge of light in small telescopes.

The Hyades is a cluster of middle-aged stars spanning nearly 5°, lying 140 light-years distant. The cluster looks its best to the naked eye; magnification destroys its unity. Aldebaran is not part of the cluster; it just happens to lie in line with it.

More spectacular is the Pleiades, a cluster of young, bright stars. To the naked eye it's a soft glow; with binoculars, it sparkles. With a small telescope, this most impressive of open clusters positively scintillates.

Gemini, the Twins

Gemini follows Taurus in the zodiac, and is a rewarding constellation for the telescope. Gemini's "feet" lie in a rich, starry region of the Milky Way.

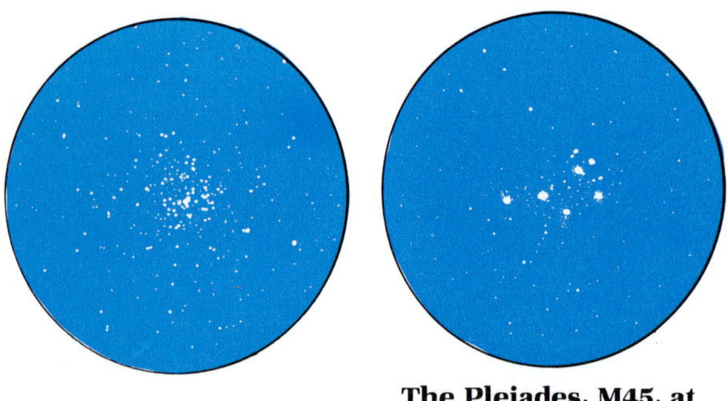

M37

The Pleiades, M45, at low power

Castor, alpha (α) Geminorum, is a binary with a period of 500 years. The brighter star is second magnitude; the companion star is third magnitude. During the 1980s and the decades following, the dim star will slowly move away from the primary.

While you're looking at doubles, why not compare delta (δ), eta (η), and lambda (λ)? Delta's primary star is magnitude 3.5, the secondary is 8.5, and they are separated by 6 seconds of arc. Lambda has a larger magnitude difference (the primary is 3.6, the secondary 10.7), but with a separation of 9.6 seconds of arc. For a real challenge try eta. The primary is a third-magnitude variable and the secondary is magnitude 8.8. They are separated by a mere 1.6 seconds of arc.

Forming the northern apex of a shallow triangle with eta and 1 Geminorum is the open cluster M35. Under clear, dark skies, this cluster is bright enough to see with the unaided eye.

Search for the Eskimo nebula, NGC 2392, 2° south and west of the double star delta. This tenth-magnitude planetary looks about the same size as Jupiter through a telescope, but it's much dimmer.

Auriga, the Charioteer

Auriga lies in a starry region of the Milky Way. Renowned for its trio of sparkling star clusters, M36, M37, and M38, Auriga is worth scanning with binoculars or a telescope for no more than the pleasure of seeing so many stars in the field of view at one time.

Of the three clusters, M37 is brightest, then comes M36, and last, M38. All are easily located with binoculars. With a telescope, they are interesting objects to compare. M37 is one of the finest clusters in the sky: it lies in a rich, starry field, and consists of a uniform sprinkling of several hundred stars. M36 has fewer stars and is more concentrated toward the center. The stars of M38 are spread out, and it looks dimmer than the other two and irregular in shape.

Epsilon (ϵ) Aurigae is an eclipsing binary star; every 27 years, it drops from 3.0 to 3.8 magnitude for two years. The last eclipse occurred in 1983–84.

The stars in star chart 1 pass overhead at:

October 23 at 4:30 A.M.; November 7 at 2:30 A.M.; November 22 at 1:30 A.M.; December 7 at 12:30 A.M.; December 23 at 11:30 P.M.; January 7 at 10:30 P.M.; January 23 at 9:30 P.M.; February 7 at 8:30 P.M.; February 21 at 7:30 P.M.

N

CAMELOPARDALIS

δ Aur

Algenib α

PERSEUS

β Algol

Capella

AURIGA

California Nebula

M38

M36

M37

Castor

Pollux

GEMINI

Alnath

The Pleiades

NGC 2392

M35

NGC 2158

ECLIPTIC

M1

E

W

TAURUS

The Hyades

Aldebaran

30

NGC 2264

NGC 2261

13

Betelgeuse

CANIS MINOR

Procyon

Bellatrix

NGC 2244 and NGC 2237

18

ORION

NGC 2301

M78

MONOCEROS

NGC 2024

ERIDANUS

M42—Orion Nebula

Rigel

M50

Saiph

R

NGC 1535

LEPUS

Sirius

S

STAR CHART 2
Overhead, facing south

Without question, Orion is the most striking constellation in the sky. Mighty Orion was the mythical hunter who boasted that he could slay any creature, only to be fatally stung on the foot by a scorpion. According to legend, the gods placed Orion in the sky opposite the scorpion so that it would never sting him again.

Orion, the Hunter

Orion is unmistakable. A brilliant quadrilateral of the stars alpha (α), beta (β), gamma (γ), and kappa (κ) marks the hunter's shoulders and knees; the stars of his belt and his sword lie within it. Lambda (λ), phi^1 (ϕ^1) and phi^2 (ϕ^2) mark his head. He faces the charging bull, holding a lionskin shield made of the six stars named pi (π) and an upraised club extending from alpha (Betelgeuse), ending in chi^1 (χ^1) and chi^2 (χ^2).

Begin your night's stargazing with some of Orion's magnificent doubles while your eyes "get used to the dark." The most impressive is Rigel, or beta (β) Orionis. Rigel is not a close double, but its companion is a small blue star. On nights of good seeing, you will view the secondary close beside blindingly bright blue-white Rigel as a faint speck of light. On poor nights, though, scattered light from Rigel will obscure its 400-times-dimmer companion.

Below is a list of doubles in Orion. Why not try them all?

Star	*Primary*	*Secondary*	*Separation*
Delta (δ)	2.2	13.7	32.8″
Iota (ι)	2.8	6.9	11.3″
Beta (β)	0.1	6.8	9.5″
Rho (ρ)	4.5	8.3	7.0″
Lambda (λ)	3.6	5.5	4.4″
Zeta (ζ)	1.9	4.0	2.3″
Eta (η)	3.8	4.8	1.5″

Orion's greatest glory is the Orion nebula. This forty-second item on Charles Messier's list of nebulous objects is impressive with practically any telescope. Even with a binocular, the nebula is a milky glow enveloping the stars in Orion's sword. With a large telescope at high magnification, you can study the intricate swirls of the nebula all night.

At the very heart of the nebula is the most famous multiple star of all, the Trapezium, theta^1

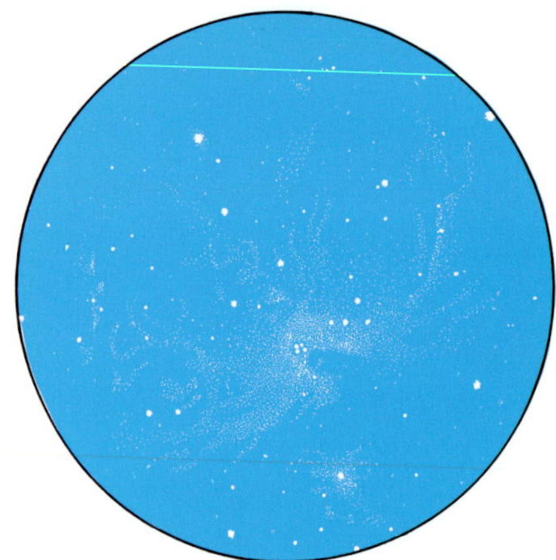

The Orion nebula, M42

(θ^1) Orionis. The stars shine at 5.1, 6.7, 6.7, and 7.9 magnitude, arranged in the trapezoid that gives the object its name. On exceptionally clear and steady nights, large telescopes may reveal two eleventh-magnitude stars, making the theta^1 system a sextuple star, at least. The dark bay in the nebula, actually dark material blocking light from the glowing gas behind it, is known as the "fish mouth." The chart shows only M42, but just north of M42 is an extension known as M43. Half a degree farther north, enveloping stars 42 and 45 Orionis, is a much fainter nebula, NGC 1977. To the south, M42 billows toward iota (ι); one nebular strand called NGC 1980 even wraps around that star.

Near zeta (ζ) are three more demanding objects. Immediately to the east of this easternmost belt star is NGC 2024; a reflection nebula that looks as if it's being ripped to shreds. Two degrees north and 1° east lies M78, a faint scrap of dust and gas illuminated by a star embedded in it. Extending south from zeta and beside sigma (σ), lies a very, very faint nebular strip known as IC 434. (This strip is so faint it's not even marked on star chart 2.) About midway down it, half a degree south of zeta, is a small dark notch caused by dark material in front of IC 434. That little notch is the famous Horsehead nebula. It is seldom seen except under the darkest skies by skilled observers.

The stars in star chart 2 are highest at:

October 7 at 5:30 A.M.; October 23 at 4:30 A.M.; November 7 at 2:30 A.M.; November 23 at 1:30 A.M.; December 7 at 12:30 A.M.; December 23 at 11:30 P.M.; January 7 at 10:30 P.M.; January 23 at 9:30 P.M.; February 7 at 8:30 P.M.; February 21 at 7:30 P.M.; March 7 at 6:30 P.M.

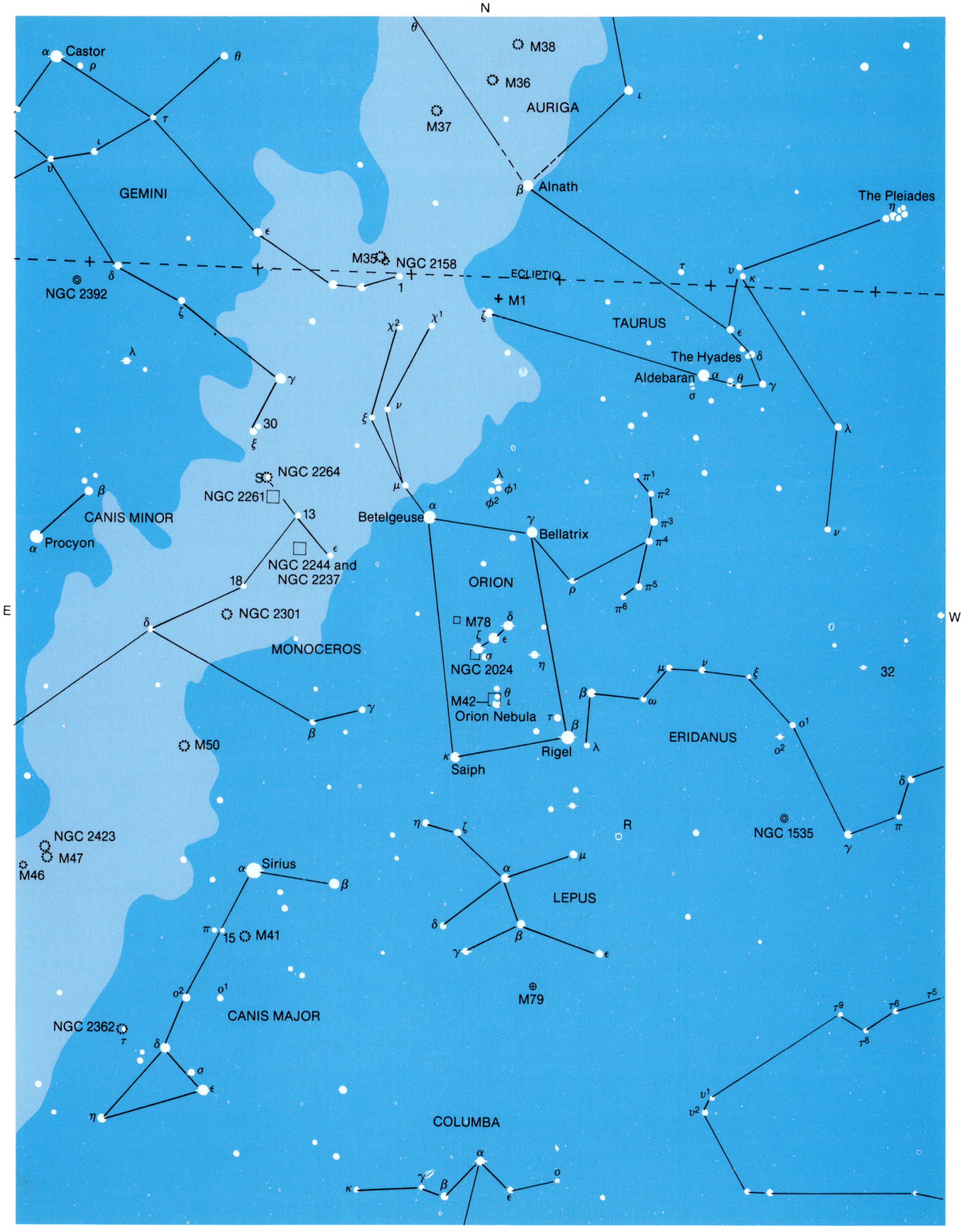

N
S
E
W
Castor
M38
M36
AURIGA
M37
GEMINI
Alnath
The Pleiades
M35
NGC 2158
ECLIPTIC
NGC 2392
M1
TAURUS
The Hyades
Aldebaran
30
NGC 2264
NGC 2261
CANIS MINOR
13
Betelgeuse
Bellatrix
Procyon
NGC 2244 and
NGC 2237
ORION
18
NGC 2301
M78
MONOCEROS
NGC 2024
32
M42
Orion Nebula
ERIDANUS
M50
Rigel
Saiph
R
NGC 2423
NGC 1535
M47
M46
Sirius
LEPUS
15
M41
M79
CANIS MAJOR
NGC 2362
COLUMBA

STAR CHART 3
Facing south

South of bright Orion is a rich region of the sky. At Orion's heel romps Canis Major, the Greater Dog, perhaps chasing Lepus, the Hare, toward the River Eridanus. Left of and below Orion shines Sirius, the bright star marking the eye of the Greater Dog, and the brightest star in the sky.

Canis Major, the Greater Dog

Sirius is the brightest star in the sky, with a faint white-dwarf companion star that orbits it in fifty years. The 10,000-to-1 difference in brightness makes the companion difficult to split with amateur telescopes even when the stars are widely separated—but they'll be at their closest in 1994! Sirius is magnitude –1.5 and the white-dwarf star companion shines at 8.5. In 1994, they will be only 3 arcseconds apart.

Adhara, epsilon (ϵ) Canis Majoris, is much easier to split than Sirius. With a 250-fold difference in their brightnesses and a separation of 7.5 arcseconds, Adhara is still a challenge for any telescope. Give it a try, and see what you can do.

The Milky Way, where it runs through Canis Major, is rich in open star clusters. M41 is one of the most beautiful, visible as a faint glow below Sirius to the naked eye, a dusting of stars through binoculars, and a showy, bright cluster through a telescope. M41 is roughly 2,000 light-years from us in space. Enveloping tau (τ) is another cluster, NGC 2362. Smaller than M41, it also lies twice as far from us.

Lepus, the Hare

Do you see the stars of Lepus as the square-faced head of a hare with mu (μ) and epsilon (ϵ) serving as its ears, or as a hare's pelt spread-eagled to cure? In either case, this little constellation at Orion's feet is an ancient one.

Gamma (γ) and kappa (κ) are both doubles, but gamma is wide and kappa is close. Gamma's third- and fourth-magnitude components lie 96 arcseconds apart, and so are readily split with a binocular. A third, eleventh-magnitude component lies 45 arcseconds farther away. It's a difficult one. Can you see it? Kappa's component stars lie 2.6 arcseconds apart, and shine at 4.5 and 7.4 magnitude. Good optics and good seeing are a must for this one.

R Leporis is a deep-orange star that skirts naked-eye visibility, then drops beyond the limit of small amateur telescopes. Called Hind's Crimson star, R is a long-period Mira-type variable with a period of 432 days. When you look at Lepus with binoculars, note whether R is bright enough to see.

To find the globular cluster M79, extend a line from alpha (α) to beta (β) and continue on an equal distance. You'll see a fifth-magnitude star and, beside it, M79. This magnitude-8 globular lies 40,000 light-years from earth. This star is a triple, the two bright components separated by 3 arcseconds, and a dim star 59 arcseconds distant.

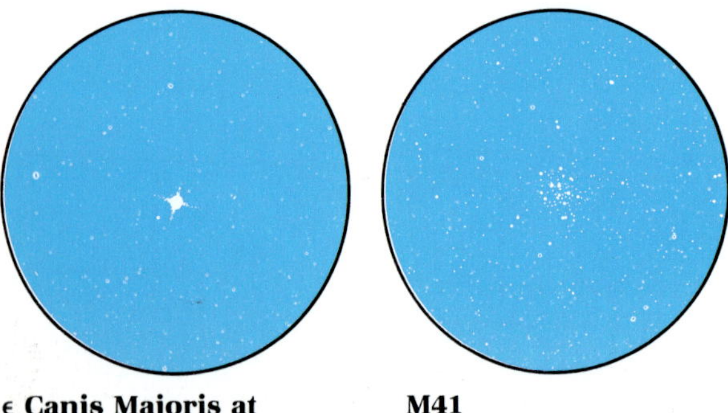

ϵ Canis Majoris at high magnification **M41**

Columba, the Dove

Columba lies south of Lepus. Observers in the southern states may wish to try for NGC 1851, a seventh-magnitude globular. Alpha (α) has a faint companion 13 arcseconds distant.

Eridanus, the River (see also star chart 23)

In eastern Eridanus look for the double 32 Eridani and a faint planetary nebula, NGC 1535. Star 32 offers a nice color contrast with its fifth-magnitude yellow primary and sixth-magnitude blue secondary. The stars are separated by 6.8 arcseconds. NGC 1535 lies almost equidistant from omicron-2 (o^2) and gamma (γ), at the apex of a right triangle. It is small and faint.

The stars in star chart 3 are highest at:

October 7 at 5:30 A.M.; October 23 at 4:30 A.M.; November 7 at 2:30 A.M.; November 23 at 1:30 A.M.; December 7 at 12:30 A.M.; December 23 at 11:30 P.M.; January 7 at 10:30 P.M.; January 23 at 9:30 P.M.; February 7 at 8:30 P.M.; February 21 at 7:30 P.M.; March 7 at 6:30 P.M.

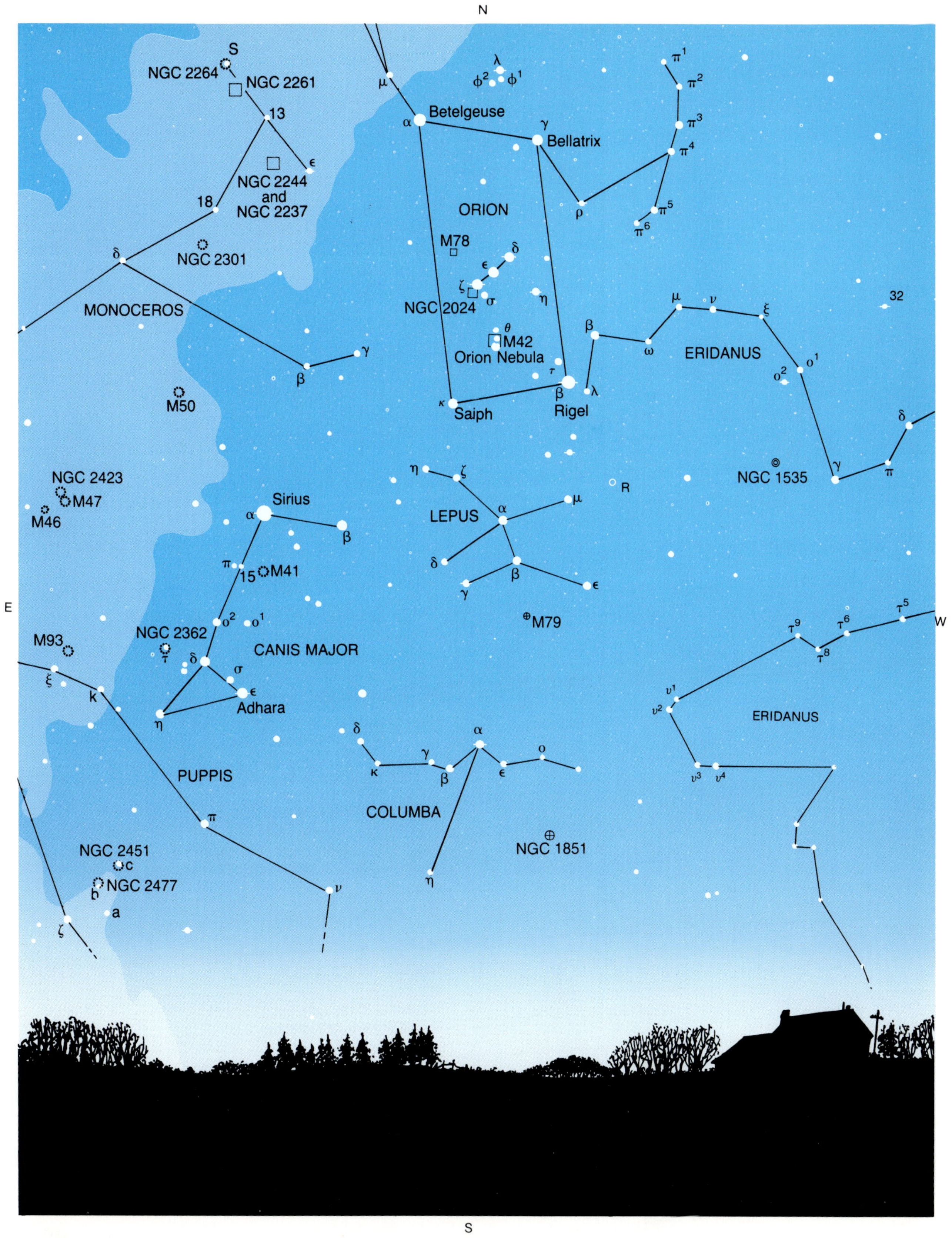

N
S
E
W
ORION
Betelgeuse
Bellatrix
Saiph
Rigel
M78
NGC 2024
M42
Orion Nebula
MONOCEROS
NGC 2264
NGC 2261
NGC 2244
and
NGC 2237
NGC 2301
M50
NGC 2423
M47
M46
ERIDANUS
NGC 1535
32
R
LEPUS
M79
Sirius
M41
CANIS MAJOR
NGC 2362
Adhara
M93
PUPPIS
NGC 2451
NGC 2477
COLUMBA
NGC 1851
ERIDANUS

STAR CHART 4
Facing north

North of the bright stars of Gemini is one of the darkest areas in the entire sky. So star-poor is this region that the constellations featured in this chart, Lynx and Camelopardalis, were invented in the seventeenth century to fill the gaping hole in star charts. Despite its dullness to the unaided eye, the region offers interesting sights for the celestial explorer with a telescope.

Camelopardalis, the Giraffe

Begin your telescopic exploration of the Giraffe at beta (β), the brightest star in the constellation. Beta is a triple. The primary is pale yellow and magnitude 4.0; the secondary is blue and magnitude 8.6 at 80 seconds of arc. Located 15 arcseconds from the secondary is the magnitude-11 tertiary.

The end of Camelopardalis near the Milky Way contains several bright star clusters. At the other end, away from the galactic dust that so often blocks objects beyond, lies a moderately bright galaxy, NGC 2403. Imagine a line from gamma (γ) Camelopardalis to omicron (o) Ursae Majoris, the star at the tip of the Great Bear's nose, then point your telescope to a spot three-fourths of the way from gamma. Sweep at low power to locate this eighth-magnitude galaxy, visible as a softly glowing oval. NGC 2403 is a loose spiral similar to the famous Triangulum galaxy, M33 (shown on star chart 22), but three times as distant. Don't forget to look at M81 and M82 while you're in the area.

Sweep for the clusters in Camelopardalis with binoculars before trying to find them with a telescope. Stock 23 is the easier of the two. Half the diameter of the moon, it appears as a spangle of stars against a background of Milky Way stars. NGC 1502 is half as large and appears as a faint glow of light. A telescope resolves it into a compact star cluster.

Lynx, the Lynx

Starmapper Johannes Hevelius named this new constellation Lynx because only an observer with "the eyes of a lynx" could see its faint shape. Lynx has only one bright star to its credit, and that star, alpha (α), is the only star in Lynx with a Greek letter designation. All the rest bear Flamsteed numbers.

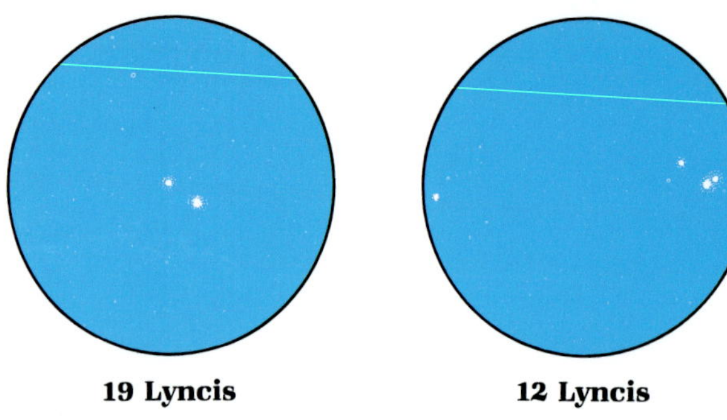

19 Lyncis **12 Lyncis**

Star 19 Lyncis is an easy double. Its 5.6- and 6.5-magnitude stars are separated by 15 seconds of arc. The two stars can be split with a large binocular, provided it is solidly mounted. The secondary looks bluish relative to the primary.

The star 38 Lyncis is tougher. The secondary shines at magnitude 6.6 against 3.9 magnitude for the primary, and is only 2.7 seconds of arc away. The pair is not easy because the secondary is so much fainter than the primary.

Quadruple star 12 is not only beautiful but also affords you the opportunity to see change in a long-period star during your lifetime. The primary star is magnitude 5.4, orbited by a magnitude 6.0 companion at a distance of 1.7 seconds of arc. Although separation is constant, the secondary is moving around the primary at a rate of 1° per year. The tertiary, magnitude 8.5 and 8.5 arcseconds away, is a good point of reference. The fourth star, a dim one, lies 170 arcseconds distant. Careful sketches of the system will show the secondary's motion in a few decades.

The star 15 Lyncis poses a real challenge. The components, magnitude 4.8 and 5.9, are separated by only 0.9 seconds of arc. Splitting this star requires steady seeing, fine telescope optics, and high magnification.

If you'd like another challenge, try to find the galaxy NGC 2683. With a medium-sized telescope, this tenth-magnitude spiral looks like the Andromeda galaxy does through binoculars.

The stars in star chart 4 are highest at:

November 7 at 4 A.M.; November 23 at 3 A.M.; December 7 at 2 A.M.; December 23 at 1 A.M.; January 7 at midnight; January 23 at 11 P.M.; February 7 at 10 P.M.; February 21 at 9 P.M.; March 7 at 8 P.M.

S
N
W
E
β Tau
M37
M36
M38
AURIGA
ι
θ
β
ζ
η
ε
α
Capella
Pollux
GEMINI
σ
τ
ρ
Castor
NGC 2683
38
31
LYNX
10 UMa
21
19
15
2
12
7
CAMELOPARDALIS
NGC 1502
Stock 23
IC 1805
NGC 663
NGC 654
M103
CASSIOPEIA
NGC 2403
URSA MAJOR
26
23
M81
M82
M108
Dubhe
DRACO
Polaris
+ NCP
URSA MINOR
Nova 1572
CEPHEUS
Thuban
10

STAR CHART 5 Overhead, facing south

With the melting of winter snows, dim Cancer and Orion's Little Dog, Canis Minor, cross the meridian. Lying between the Twins and the Lion, Cancer is the least conspicuous of the constellations of the zodiac. Although the Little Dog's luminary Procyon is a facet of the brilliant Winter Hexagon, this region is dim by comparison to the bright stars to the east.

Cancer, the Crab

Cancer is a dim constellation, but it is rich in deep-sky objects. Most famous is the bright star cluster called the Beehive, easily located with binoculars, but observers should not overlook a dimmer cluster and several challenging double stars.

The Beehive Cluster (M44) lies centrally on the Crab's carapace, over 1° across, visible to the naked eye on a clear night. With binoculars it is a magnificent spangle of stars, many seeming to form pairs and triplets. When viewing the Beehive with a telescope, use the lowest magnification you have. By contrast, the star cluster M67 is dimmer and smaller than the Beehive. This star cluster lies 1° west of alpha (α) Cancri.

Iota (ι) is an easy double—fourth- and sixth-magnitude stars lying 31 arcseconds apart. If you can hold a binocular steady enough, you'll see this as a double star. With a telescope, note the contrasting yellow and blue colors of the stars.

Zeta (ζ) is a fine and challenging quadruple star. The primary is magnitude 5.6, and very, very close to it—at only 0.6 arcseconds—orbits the sixth-magnitude secondary, with a period of 60 years. Many small telescopes won't be able to split this pair at all. However, circling them at some 5.9 arcseconds is a 6.2-magnitude tertiary. It orbits the pair in 1,150 years. Some 280 arcseconds away, a ninth-magnitude quaternary completes this remarkable system.

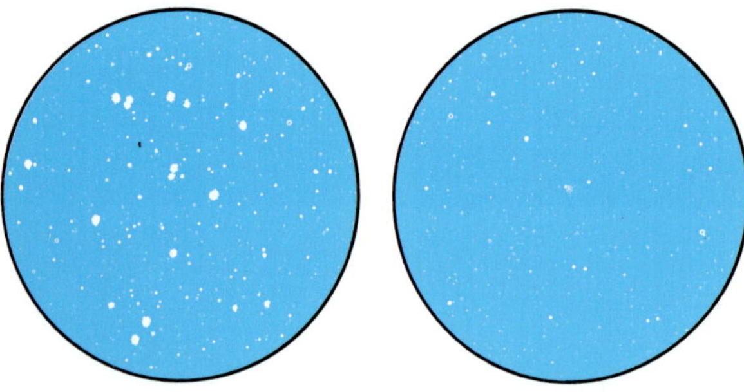

The Beehive Cluster, M44

NGC 2261, Hubble's variable nebula

Canis Minor, the Little Dog

Tiny Canis Minor has little to offer but its one brilliant star. Procyon, alpha (α) Canis Minoris, is near us in space—11 light-years away—and is one of the brightest stars in the sky. It has a tenth-magnitude companion that orbits it every 41 years, a bit over 5 arcseconds from its ten-thousand-times brighter primary. Looking for such a faint, close companion is futile without a large-aperture observatory telescope.

Monoceros, the Unicorn (see also star chart 2)

Lying astride the winter Milky Way, the Unicorn is rich in star clusters and nebulae. Search them out on a dark, clear night. Particularly interesting are NGC 2244, the star cluster embedded in the Rosette nebula, and NGC 2264, the cluster enveloped in the Cone nebula. The Rosette, NGC 2237, shows up on long-exposure photographs as a ruddy doughnut over 1° in diameter, and is visible in telescopes as a faint haze surrounding the cluster. The Cone is considerably fainter and more difficult to see.

A famous though faint attraction in this area is NGC 2261, Hubble's variable nebula. This triangular patch of dust and gas reflects the light of a variable star embedded in it. The nebula is faint—usually around tenth magnitude—and only 1 by 5 arcminutes across.

You may wish to end an evening's exploration with the triple star beta (β) (shown on star chart 2). The 4.7 magnitude primary has two companions: one, a fifth-magnitude star 7 arcseconds from it and the other a sixth-magnitude 10 arcseconds away. Epsilon (ϵ) is a pretty double consisting of 4.5- and 6.5-magnitude stars separated by 14 arcseconds.

The stars in star chart 5 pass overhead at:

November 7 at 6 A.M.; November 23 at 5 A.M.; December 7 at 4 A.M.; December 23 at 3 A.M.; January 7 at 2 A.M.; January 23 at 1 A.M.; February 7 at midnight; February 21 at 11 P.M.; March 7 at 10 P.M.; March 23 at 9 P.M.; April 7 at 9 P.M.; April 22 at 8 P.M.

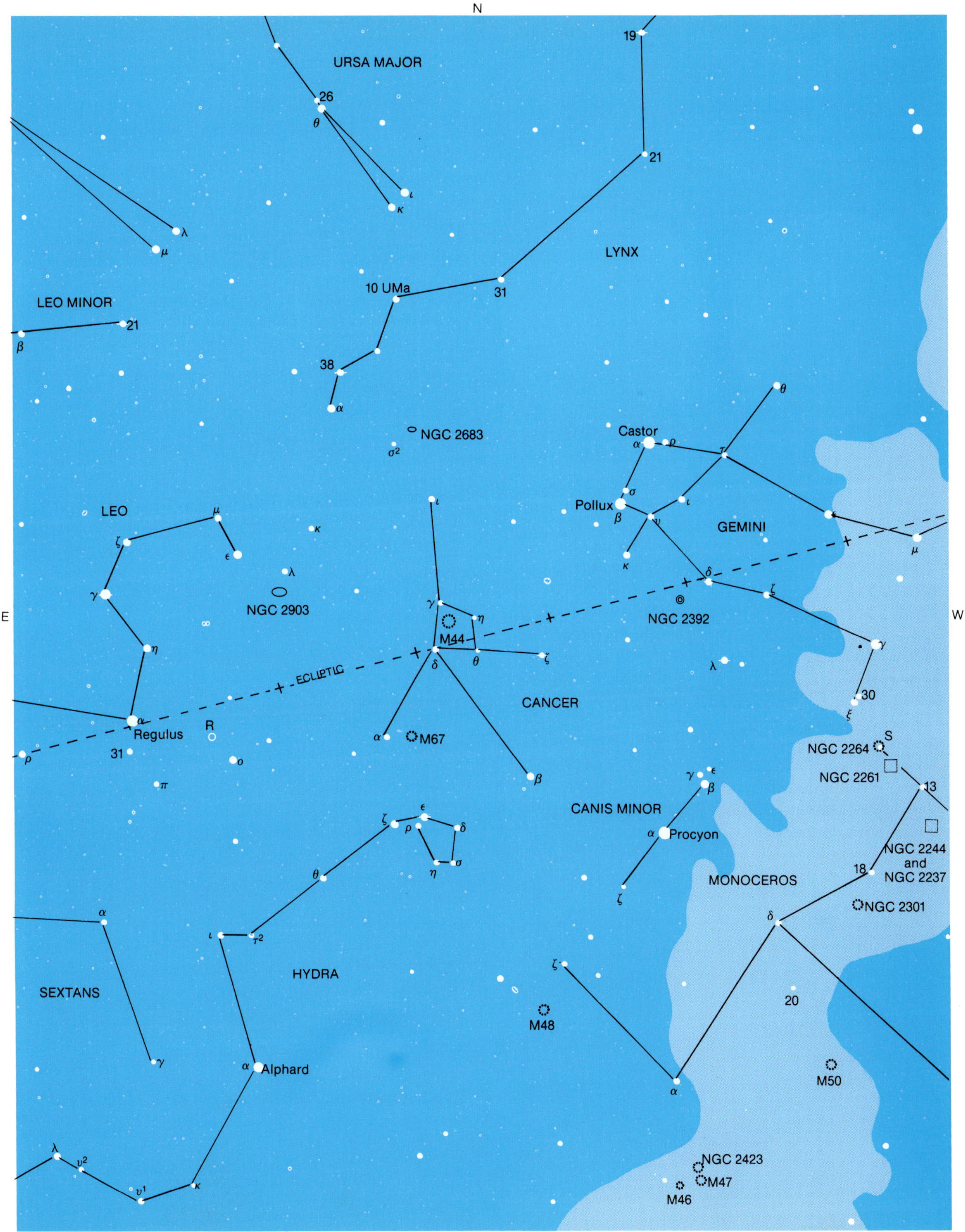

N
URSA MAJOR
LYNX
LEO MINOR
10 UMa
NGC 2683
Castor
Pollux
GEMINI
LEO
NGC 2903
M44
ECLIPTIC
NGC 2392
Regulus
R
M67
CANCER
CANIS MINOR
Procyon
NGC 2264
NGC 2261
NGC 2244
and
NGC 2237
MONOCEROS
NGC 2301
SEXTANS
HYDRA
M48
Alphard
M50
NGC 2423
M47
M46
E
W
S

STAR CHART 6
Facing south

South of Gemini and Leo, the winter Milky Way plunges into the horizon amid the stars that once made up Argo, named for the ship brave Jason sailed in his search for the Golden Fleece. The ship sails the sky stern-first. Puppis is the poop, or sterncastle. Vela is the sails, and faint Pyxis the ship's compass. (Below the horizon is Carina, the keel.) This chart shows a rich and star-filled region, a delight to binocular- and telescope-user alike.

Puppis, the Stern

Puppis, amid the rich starfields of the Milky Way, abounds in open clusters. M47, in the northern part of Puppis, is naked-eye bright and quite striking through a binocular, its three dozen stars covering an area the size of the full moon. Adjacent to it are two more, M46 and NGC 2423, clusters with more numerous but dimmer stars than M47.

With a telescope, M46 reveals a striking feature. Look carefully at it. Among its hundred or so stars you'll see that one is a planetary nebula. It's tenth magnitude, and approximately 60 arcseconds across. You'll find another cluster, M93, 1° north and west of xi (ξ) Puppis. Through a telescope at low magnification it's a pretty object.

Farther south, in the glittery section of the Milky Way near zeta (ζ) Puppis, one of the bluest stars in the sky, are two more open clusters. NGC 2477 is the smaller, dimmer, and starrier of the two. NGC 2451 has a quarter as many stars as NGC2477, but is much brighter because it is only a fourth as distant (850 light-years). They're an interesting couple of clusters to compare.

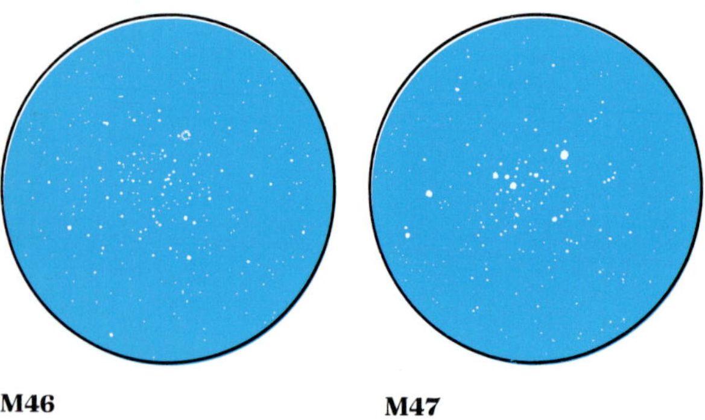

M46 M47

Star k Puppis is a striking double composed of two almost equally bright blue white stars 10 arcseconds apart. In contrast, sigma's (σ) primary is six magnitudes brighter than a ninth-magnitude companion star 22 arcseconds distant. For North American observers, sigma never rises high in the sky.

Vela, the Sails

From North America, Vela's rich Milky Way fields are either sadly dimmed by atmospheric absorption or below the horizon entirely. If the primary star is high enough, take a look at the quadruple star gamma (γ) Velorum. The primary is a superhot blue-white star, the brightest such known. Nearby you'll see a fourth-magnitude companion star 41 arcseconds away, and magnitude 8 and 9 companions at 62 and 93 arcseconds, respectively.

Pyxis, Ship's Compass

Insignificant as a constellation can be, tiny Pyxis has one interesting feature to show: T Pyxidis, a recurrent nova. Most of the time, the star is fourteenth magnitude—too dim for any but a large telescope. Roughly every 20 years, though, it flares to sixth magnitude. It erupted in 1966; the next brightening could occur at anytime.

Hydra, the Water Snake (see also star charts 8 and 12)

East of the Milky Way lies a great dull expanse of sky, and across that gulf winds Hydra, the Water Snake. Alpha (α) Hydrae, whose name Alphard means "the solitary one," is the brightest star in the region. Barely within Hydra's western boundary is M48, an open cluster almost 1° across, an easy object for a binocular.

In the Snake's head is a triple star, epsilon (ϵ), that appears as a double in all but the largest instruments. The "primary" is actually two stars separated by 0.2 arcseconds. They orbit each other in 15 years. The seventh-magnitude tertiary star lies 6.8 arcseconds from this pair, easily resolved with small telescopes.

The stars in star chart 6 are highest at:

November 7 at 6 A.M.; November 22 at 5 A.M.; December 7 at 4 A.M.; December 23 at 3 A.M.; January 7 at 2 A.M.; January 23 at 1 A.M.; February 7 at midnight; February 21 at 11 P.M.; March 7 at 10 P.M.; March 23 at 9 P.M.; April 7 at 9 P.M.; April 22 at 8 P.M.

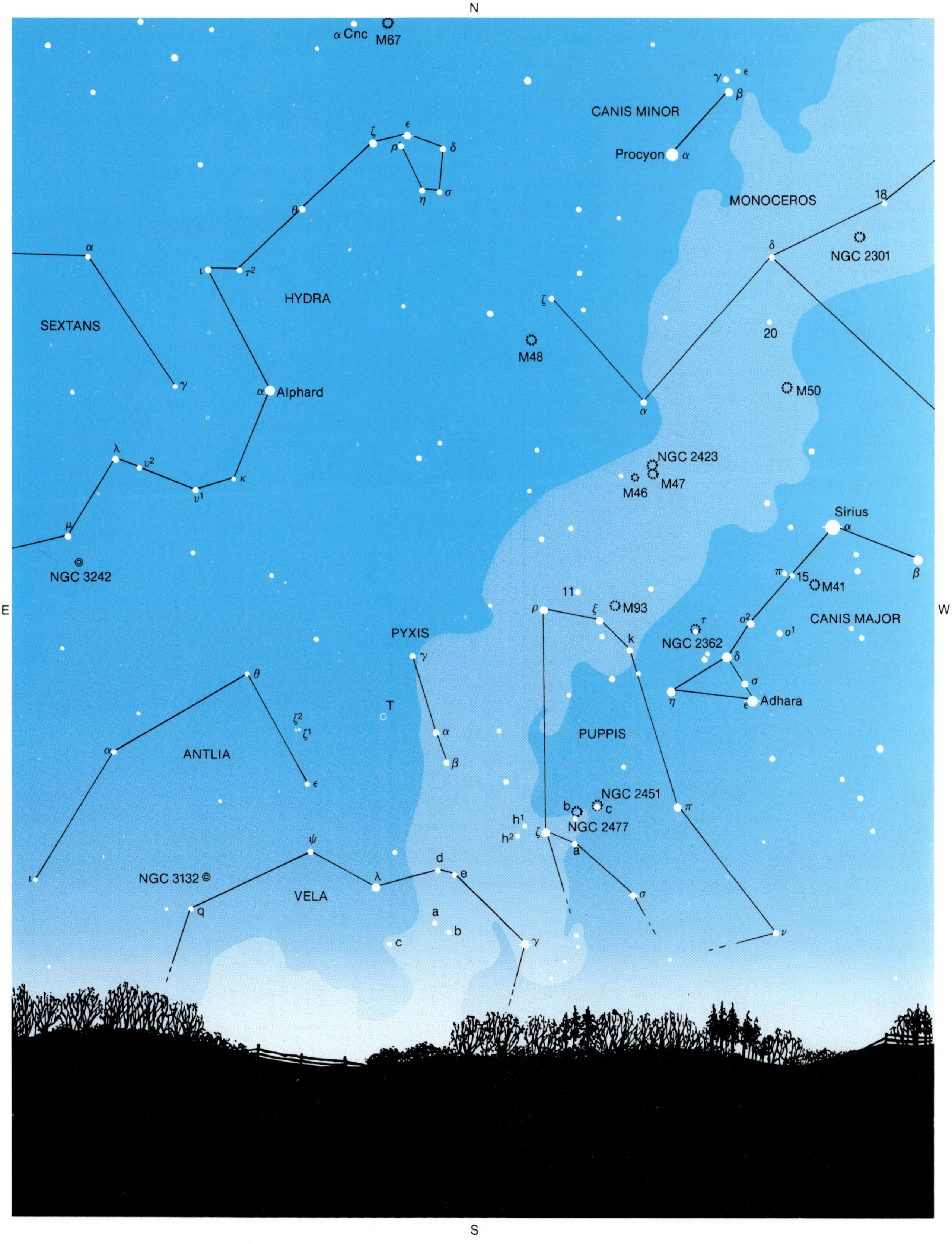

N

α Cnc
M67

CANIS MINOR
Procyon

MONOCEROS
18
NGC 2301
20
M50

HYDRA
SEXTANS
Alphard
M48

NGC 2423
M47
M46

NGC 3242

Sirius
15
M41
CANIS MAJOR

E
W

11
M93
PYXIS
NGC 2362
Adhara

T
ANTLIA
PUPPIS

NGC 2451
NGC 2477
h^1
h^2

NGC 3132
VELA

S

STAR CHART 7
Overhead, facing south

South of the seven bright stars that mark both Ursa Major, the Great Bear, and the Big Dipper, crouches Leo, the Lion. While the king of beasts guards his section of the ecliptic in regal dignity, the Lesser Lion frolics under the Bear's hind paws! Leo is a bright constellation, easy to spot and a useful guide in a region largely devoid of bright stars.

Leo, the Lion

Harbinger of spring, Leo rises as the bright constellations of winter set. Though it is not correctly placed in the stick-figure Lion, Regulus is said to mark his heart. With binoculars or a small telescope, this bright star is a wide double. The seventh-magnitude companion lies 176 arcseconds—just under 3 arcminutes—from its 1.4-magnitude primary.

Gamma (γ) is a bright binary. Separated by 4.3 arcseconds, both components, one 2.2 and the other 3.5 magnitude, have a tawny cast. Nearby fifth-magnitude 40 Leonis forms a pretty optical double with gamma.

R Leonis is a bright long-period variable star, located conveniently near Regulus. At its brightest, it shines at fourth magnitude but drops to eleventh magnitude at minimum. Its period of variation is 312 days. Check on R Leonis each time you look at Leo, and write in your observing notebook whether you can see this variable.

Leo hosts several noteworthy galaxies. Below the hindquarters of the Lion is the pair M65 and M66, just about equidistant from theta (θ) and iota (ι) Leonis. Both galaxies are spirals seen at a fairly steep angle. With a small telescope, they appear as two ovals half a degree apart. A faint galaxy, NGC 3628, lies near them.

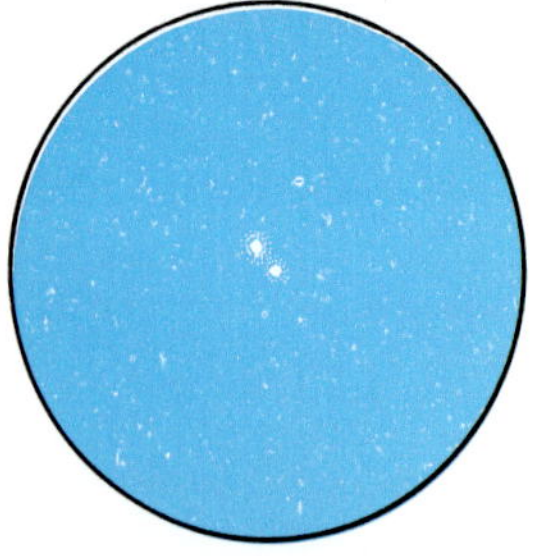

γ Leonis at high magnification

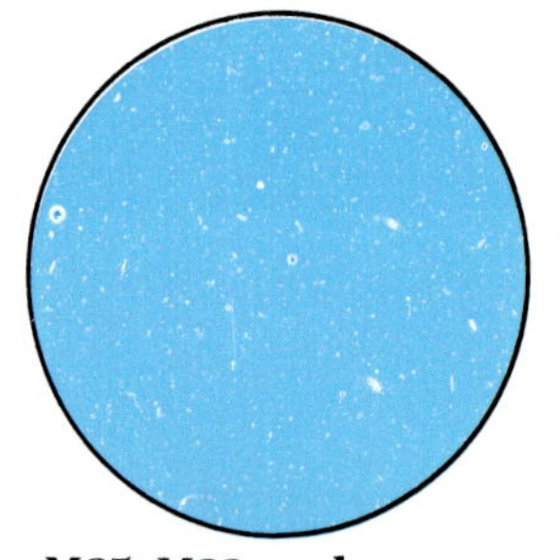

M65, M66, and NGC 3628

Messier's objects 95, 96, and 105 (called the "Leo Trio") are found just north of 53 Leonis. Through a small telescope, M95 and M96 are ninth-magnitude oval glows; nearby is M105, a circular patch of light. NGC 2903 is a large, fairly bright spiral ahead of Leo's "sickle." Use low magnification when searching for it.

Leo Minor, the Lesser Lion

Leo Minor is a faint constellation formed from a few leftover stars between Leo and Ursa Major. Beta (β) is a close binary with a period of 37 years. In 1992, at their closest, no telescope will be able to separate the primary and secondary.

Coma Berenices, Berenice's Hair

The "hair" of Coma Berenices, just south of gamma (γ), is a loose naked-eye star cluster some 250 light-years from us. With a binocular, you'll see roughly eighty cluster stars in an area about 4° across. If it were ten times as far away, Coma would resemble the many open clusters sprinkled along the Milky Way.

Coma offers a variety of deep-sky objects. Star 24 (see star chart 10) is a very pretty double with strongly contrasting stars. The fifth-magnitude primary is orangey in color; 20 arcseconds from it lies its sixth-magnitude blue-white companion. You'll find M53, a globular cluster, right next to alpha (α). (This star is known as Diadem, the gem in Berenice's hair.) Thus located, M53 is easy to find, and is fairly bright. M64 is a remarkable spiral galaxy with a dark dust band across its face. Visible with small telescopes, this odd feature has given this object its name: the Black Eye galaxy. Next, turn your telescope toward NGC 4565. This edge-on spiral galaxy looks pencil-thin through medium and large amateur telescopes. With small telescopes, it resembles a slender needle of light in the sky.

The Virgo Cluster of galaxies extends northward into Coma. These galaxies are covered in star chart 10.

The stars in star chart 7 are highest at:

December 23 at 5 A.M.; January 7 at 4 A.M.; January 23 at 3 A.M.; February 7 at 2 A.M.; February 21 at 1 A.M.; March 7 at midnight; March 23 at 11 P.M.; April 7 at 11 P.M.; April 22 at 10 P.M.; May 7 at 9 P.M.

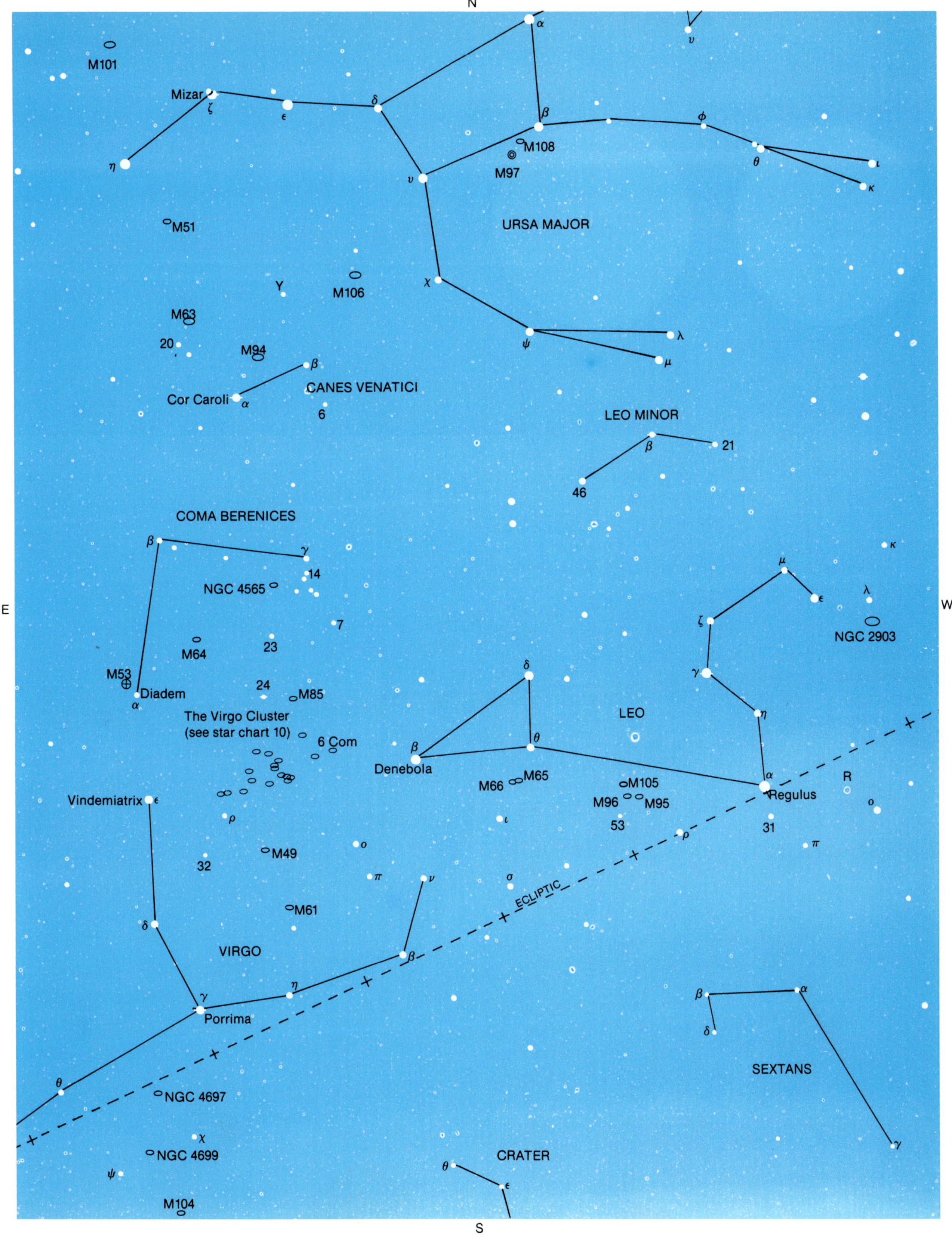
N
S
E
W
M101
Mizar
M108
M97
URSA MAJOR
M51
Y
M106
M63
20
M94
CANES VENATICI
Cor Caroli
6
LEO MINOR
21
46
COMA BERENICES
14
NGC 4565
7
NGC 2903
M64
23
M53
Diadem
24
M85
The Virgo Cluster
(see star chart 10)
6 Com
LEO
Denebola
M66
M65
M105
R
M96
M95
Regulus
Vindemiatrix
53
31
M49
32
ECLIPTIC
M61
VIRGO
Porrima
SEXTANS
NGC 4697
NGC 4699
CRATER
M104

STAR CHART 8
Facing south

South of Leo and Virgo, this relatively dark section of sky is dominated by the bright trapezoid of stars marking Corvus, the Crow. Trace the twisted coils of Hydra from the head (on star chart 6) across this chart to its tail (in star chart 12). Balanced precariously on the Snake's back is Crater, the Goblet; below it the inconspicuous southern constellation, Antlia, the Air Pump.

Corvus, the Crow

The mnemonic "Follow the arc to Arcturus, and speed to Spica, then curve to Corvus" carries you halfway across the sky to the Crow. The most prominent constellation in this rather dark, lusterless section of the spring sky, Corvus does not lie in a particularly interesting region for the telescope; we are looking neither at the rich fields of the Milky Way, nor at clusters of galaxies far beyond.

Nonetheless, whenever you have a chance, check on the variable star R Corvi. This variable is the same type as Mira in Cetus (star chart 23), rising to 6.7 magnitude and falling to 14.4 with a period of 317 days. Check its location with binoculars. R is nestled between two seventh-magnitude stars; if you see a third star between them, R is within binocular grasp, roughly ninth magnitude or brighter. Keep a record of your sightings of R in your observing notebook, and you'll see it come and go.

Third-magnitude delta (δ) is an easy double for small telescopes. Its ninth-magnitude companion lies 24 arcseconds away. How small a telescope will reveal delta's companion? Make a set of cardboard "stops" for your telescope, with 1-inch, 1.5-inch, 2-inch, and 2.5-inch openings. Try them with this and other double stars. You'll notice that the bigger the aperture of the telescope, the smaller and sharper the image of the star. And, of course, larger apertures give brighter images, too.

South and west of delta is a ninth-magnitude planetary nebula, NGC 4361. It's not an easy object even for a medium-sized telescope. Another even more challenging object lies west of gamma (γ) Corvi, a pair of galaxies called "The Antennae," NGC 4038 and 4039. These two galaxies, a binary, are pulling stars off each other. With a medium or large telescope, they appear as two close, faint eleventh-magnitude glows. It's not easy to locate this pair, so don't feel bad if you can't.

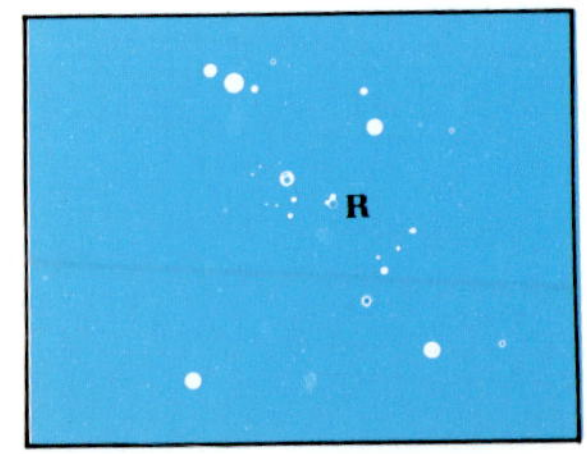

R Corvi

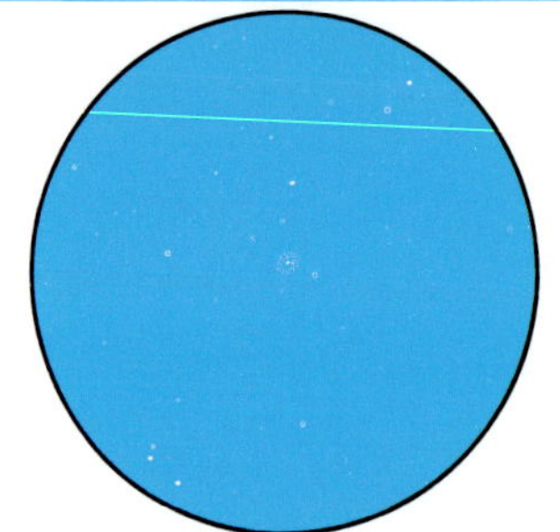

NGC 3242

Crater, the Goblet

Gamma (γ) Crateris is a tough double. The primary is 4.1 magnitude; its tenth-magnitude companion lies only 5 arcseconds away. Splitting this star is a challenge for small telescopes.

Hydra, the Water Snake (see also star charts 6 and 12)

The midsection of Hydra offers two deep-sky wonders, both eighth magnitude. The globular cluster M68 is beside a fifth-magnitude star that lies on a line from delta (δ) to beta (β) Corvi extended. With a medium-sized telescope at high magnification, you should be able to resolve M68 into individual stars.

The planetary nebula NGC 3242 goes by the nickname "the Ghost of Jupiter," and it does indeed look like a faded planet. The disk appears round and well defined with large amateur telescopes.

Antlia, the Air Pump

Zeta (ζ) is a wide double in binoculars, and zeta-1 a double for telescopes. Zeta's components zeta-1 and zeta-2 are some 200 arcseconds apart, but the zeta-1 double's stars lie only 8 arcseconds from one another. If you live in the southern states, look for the bright planetary NGC 3132, just "across the border" in Vela.

The stars in star chart 8 are highest at:

December 23 at 5 A.M.; January 7 at 4 A.M.; January 23 at 3 A.M.; February 7 at 2 A.M.; February 21 at 1 A.M.; March 7 at midnight; March 23 at 11 P.M.; April 7 at 11 P.M.; April 22 at 10 P.M.; May 7 at 9 P.M.

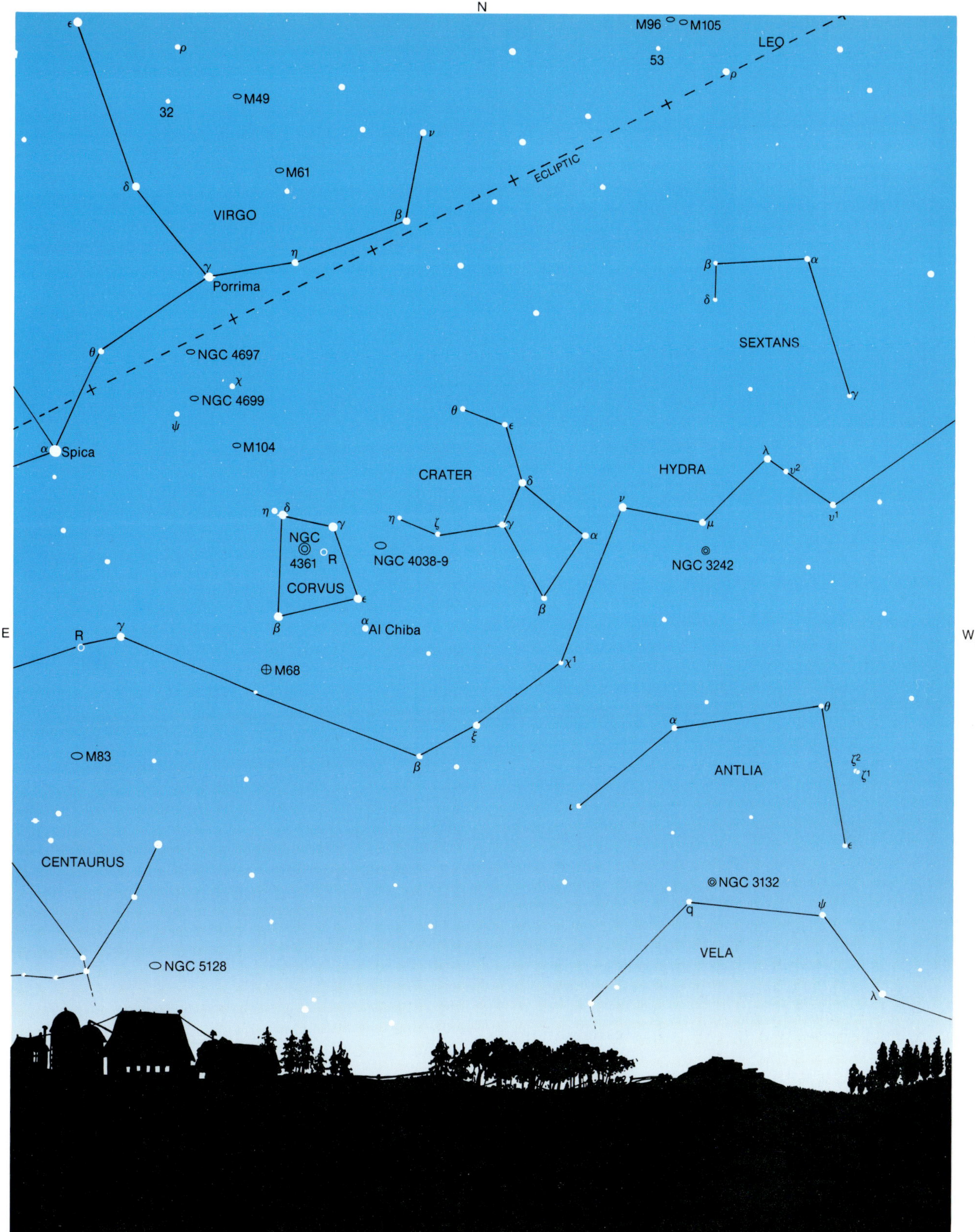

N
S
E
W
M96
M105
LEO
53
M49
32
M61
VIRGO
ECLIPTIC
Porrima
NGC 4697
NGC 4699
M104
Spica
SEXTANS
CRATER
HYDRA
NGC
4361
R
CORVUS
NGC 4038-9
NGC 3242
Al Chiba
R
M68
M83
ANTLIA
CENTAURUS
NGC 3132
VELA
NGC 5128

STAR CHART 9
Facing north

This chart shows the sky as it looks when you face north on a late spring evening when both Dippers are highest. (Of course, the Dippers are dippers in popular speech, but bears in astronomical nomenclature.) All the stars in the sky circle Polaris, the star at the tip of the Lesser Bear's tail, but both Dippers are close to the pole, so they never set.

Ursa Major, the Great Bear

One of the sky's finest doubles shines in the Big Dipper's handle. It is zeta (ζ), also known as Mizar. Sharp-eyed observers can see fourth-magnitude Alcor close upon Mizar, and binoculars certainly will suffice to separate the pair. Alcor lies 708 arcseconds from Mizar. With a telescope, examine Mizar—it's double, too. Mizar's near companion is fourth magnitude and lies 14 arcseconds from its primary.

The Great Bear is full of distant galaxies. The best two form a pair above the bear's head, about where his ears might be. M81 is a seventh-magnitude spiral, a nice oval fuzz-ball. Across a low-power field, you'll see M82, smaller and elongated. M82 is an abnormal galaxy that emits powerful radio waves. Way off at the bear's other end is M101, a face-on spiral galaxy. Binoculars and low-power telescopes offer the best views because M101 is big but dim.

Meanwhile, there's a galaxy, M108, and a planetary nebula side by side right under the bowl of the Big Dipper. From beta (β), head toward gamma (γ) and angle a bit south. M108 is smaller and dimmer than the galaxies we've already looked at, and quite elongated. The Owl nebula, M97, is small and round. The Owl and M108 are quite a challenge to new observers.

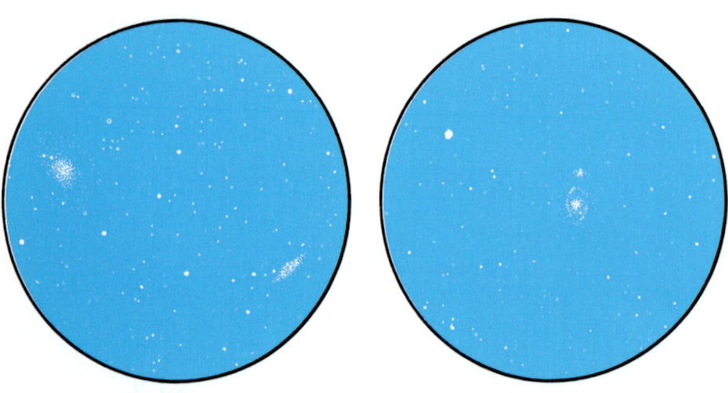

M81 and M82 **The Whirlpool galaxy, M51**

Ursa Minor, the Lesser Bear

Ursa Minor's best attraction is Polaris, the North Star. Polaris isn't exactly at the north celestial pole. It lies 44 arcminutes from the pole in the direction of the constellations Perseus and Aries. Not only is it a double, but it's also a variable, though its range of variation is small enough that it's difficult to see. Its ninth-magnitude bluish companion lies 18 arcseconds away. Polaris is a variable star of the Cepheid type (see star chart 16), and runs through a complete cycle of variation every 3 days and 23 hours.

Canes Venatici, the Hunting Dogs

This dim constellation lies between Coma Berenices (star chart 7) to the south and Ursa Major to the north. Despite its dimness, the region is interesting because it occupies the northern reaches of the Virgo Cluster of galaxies (shown on star chart 10). Of Coma's stars only Cor Caroli holds significant interest. Cor Caroli is a double; its companion is magnitude 5.5, and lies 19 arcseconds from the magnitude 2.9 primary.

M51 is the famous Whirlpool galaxy. To find it, go from zeta (ζ) and eta (η) Ursae Majoris, then take a right-angle turn toward Canes Venatici and go half the distance between zeta and eta. Because the Whirlpool has a tiny companion galaxy, you'll first see it as two faint blurs. Under a dark sky with a medium-to-large telescope, you should be able to make out traces of this galaxy's spiral arms. We see M51 almost face-on, so the arms appear in plain view.

M63 and M94 appear as softly glowing ovals, brightest in the center and tapering toward the edge. With a large telescope, you may be able to make out some detail, but their arms are not nearly so prominent as the arms of M51. M106 is longer and thinner than the others because we see it tipped at a steeper angle.

The stars in star chart 9 are highest at:

January 23 at 5 A.M.; February 7 at 4 A.M.; February 21 at 3 A.M.; March 7 at 2 A.M.; March 23 at 1 A.M.; April 7 at 1 A.M.; April 22 at midnight; May 7 at 11 P.M.; May 23 at 10 P.M.; June 7 at 9 P.M.

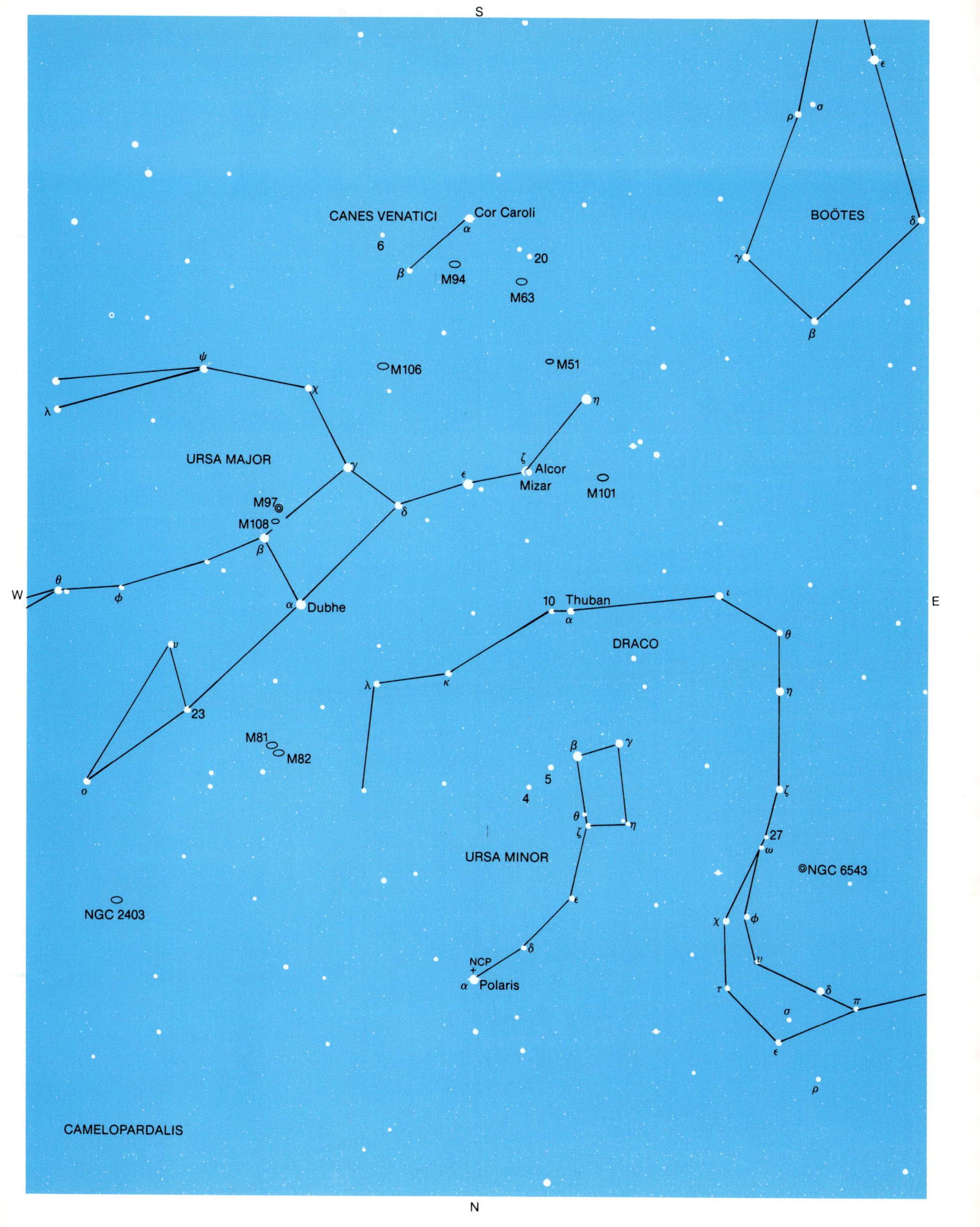

S

ε

ρ σ

CANES VENATICI

Cor Caroli

α

6

20

β

M94

M63

BOÖTES

δ

γ

β

ψ

M106

M51

χ

λ

η

URSA MAJOR

ζ

Alcor

Mizar

γ

ε

M101

M97

M108

δ

β

θ

W

φ

α

Dubhe

10

Thuban

ι

E

α

θ

DRACO

υ

λ

κ

η

23

M81

M82

β

γ

5

o

4

θ

ζ

η

ζ

27

URSA MINOR

ω

NGC 6543

NGC 2403

ε

χ

φ

δ

υ

NCP

α

Polaris

τ

δ

π

σ

ε

ρ

CAMELOPARDALIS

N

STAR CHART 10
Overhead, facing south

South of the Big Dipper and bright Arcturus, the driver of bears, Virgo, ancient goddess of fertility reclining languidly on the ecliptic, heralds the coming of crops. Virgo's brightest star, Spica, marks a stalk of wheat held in her hand.

Virgo, the Maiden

Nestled in the "bowl" of Virgo, between the stars Vindemiatrix, epsilon (ϵ) Virginis, and Denebola, beta (β) Leonis, stretches a rich cluster of galaxies, the Virgo Cluster. Some 50 million light-years from our planet, this cluster contains well over 2,000 galaxies.

With a medium-sized telescope, you'll be able to find more galaxies than are plotted in this chart. For serious galaxy-searching, though, you'll need a dark sky. If haze and city light cause you difficulty in seeing the Milky Way, then you'll also have trouble seeing these far-flung galaxies.

There are so many galaxies that your greatest problem may be figuring out which is which. One solution is to simply scan through the Virgo Cluster at low magnification, looking at the galaxies. Another solution is to start from a well-defined jumping-off point and "galaxy hop" across the cluster.

Rho (ρ) Virginis makes a good starting point, handy because you can starhop to it from delta (δ) via 32 Virginis. From rho, hop straight north to the ninth-magnitude elliptical galaxy M60 (see inset), then to tenth-magnitude M59, a little north and west. Continue west to the spiral M58, and thence to M87, a huge elliptical near the center of the cluster, and its brightest galaxy. M87 is a strong radio source. Continue north and west to the Virgo "main line," a curving arc of galaxies that starts at M84, runs through M86, and continues toward M88. At this point you leave Virgo and enter Coma Berenices. With an 8-inch or 10-inch telescope, you can see a dozen or more galaxies in a single low-power field of view.

M87

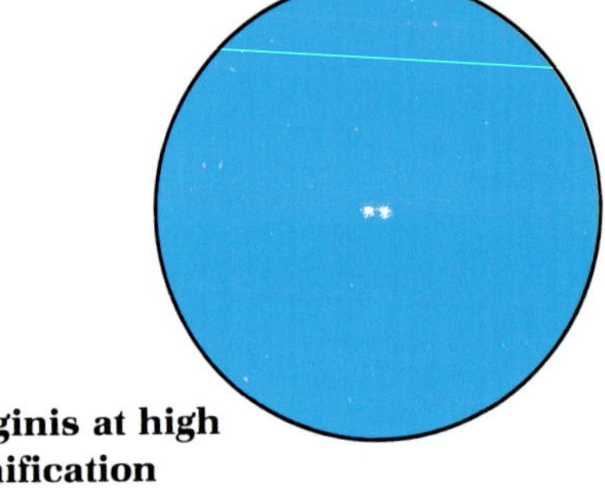

γ Virginis at high magnification

From M88, take a short detour east to pick up M91, a faint spiral galaxy; then head west to M99, a face-on spiral that looks like a miniature version of M33 (see star chart 22). Hop north and west over the star 6 Comae to M98. Hop east and north to M100, another face-on spiral, then north to M85, an elliptical.

Of course, there's nothing special about this particular route. You can make up other tours of the Virgo Cluster to suit yourself.

Another Virgo galaxy deserves special attention: M104. M104 is well off the beaten path. It is due south of gamma (γ) Virginis and due west of Spica. Known as the Sombrero galaxy, M104 is a spiral with a light-absorbing band of dust in its equator. It looks a bit like a galaxian version of Saturn, with a dark line through the middle.

In addition to hordes of galaxies, Virgo boasts a particularly interesting double star. Gamma (γ) Virginis is among the most attractive doubles in the sky. The components—both magnitude 3.5 and yellow-orange in color—are closing rapidly. In 1990, the separation between them will be 3.0 arcseconds, and in 2000, 1.8 arcseconds. Here's a star you'll come back to time after time as the years roll by. The orbital period is 171 years.

The stars in star chart 10 are highest at:

January 23 at 6 A.M.; February 7 at 5 A.M.; February 21 at 4 A.M.; March 7 at 3 A.M.; March 23 at 2 A.M.; April 7 at 2 A.M.; April 22 at 1 A.M.; May 7 at midnight; May 23 at 11 P.M.; June 7 at 10 P.M.

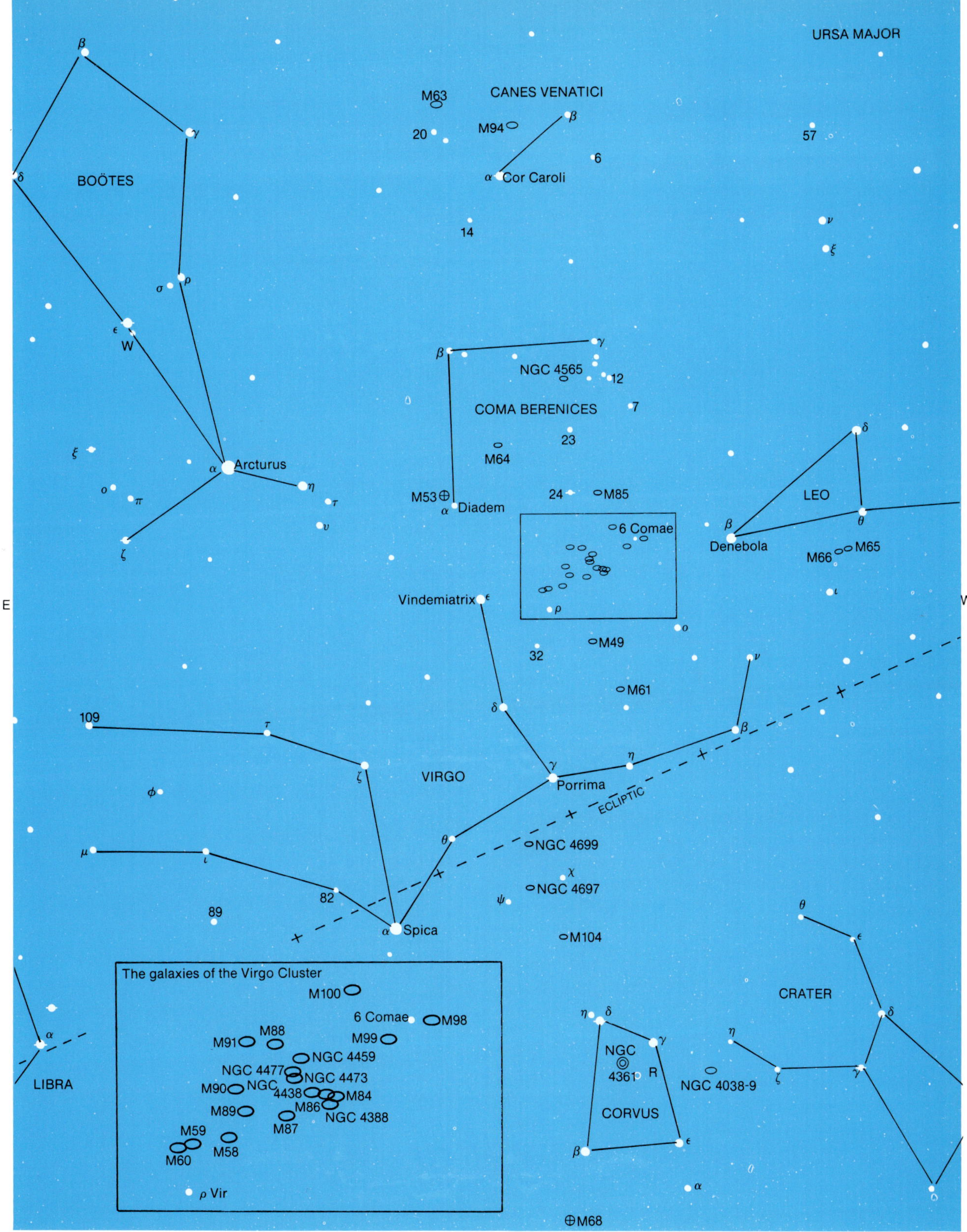

N
S
E
W
URSA MAJOR
CANES VENATICI
M63
M94
20
6
Cor Caroli
14
57
BOÖTES
W
Arcturus
COMA BERENICES
NGC 4565
12
7
23
M64
M53
Diadem
24
M85
6 Comae
LEO
Denebola
M66
M65
Vindemiatrix
M49
32
M61
109
VIRGO
Porrima
ECLIPTIC
NGC 4699
NGC 4697
82
89
Spica
M104
The galaxies of the Virgo Cluster
M100
6 Comae
M98
M91
M88
M99
NGC 4459
NGC 4477
NGC 4473
M90
NGC 4438
M84
M86
M89
NGC 4388
M87
M59
M58
M60
ρ Vir
LIBRA
CRATER
NGC 4361
R
NGC 4038-9
CORVUS
M68

STAR CHART 11
Overhead, facing south

This chart spans the heavens overhead on a late spring evening, from the end stars in the handle of the Big Dipper to Libra, low in the south. From yellow-orange Arcturus, at the center, we find our way to the unmistakable arc of the Northern Crown, the languid allure of Virgo and tangle of Berenice's Hair, and the head of the Snake which Ophiuchus (star charts 13 and 14) holds so tightly.

Boötes, the Herdsman

Turn your telescope on Arcturus for a few minutes, and savor the delicate yellow-orange of its light. It is a nearby star, the fourth brightest in the sky, and a giant with a surface that's cooler than the sun's, which is why its light appears so ruddy. After looking at Arcturus, observe the color of Vega (star chart 13) for comparison.

Boötes's finest double is Izar, epsilon (ϵ) Boötis. It is a close object with a large magnitude difference, however, so small telescopes will split it only with difficulty. The golden primary shines at magnitude 2.5 and the blue secondary at 4.9; the components lie 2.8 arcseconds apart.

Two easy objects for small telescopes are kappa (κ) and xi (ξ). Kappa's fifth- and seventh-magnitude components are 13 arcseconds distant from one another. Xi is a bit more difficult. The components are nearly the same brightness as kappa's, but the separation is only 7 arcseconds.

For a bit more challenge, try the triple star mu (μ). The yellow primary star shines at 4.3 magnitude. With binoculars, you will see a sixth-magnitude secondary 108 arcseconds from it; but with a telescope, the secondary turns out to be a binary: stars of 7.0 and 7.6 magnitude 2.3 arcseconds apart. Their orbital period is 260 years.

To challenge the optics of your telescope and your observing skill, try splitting zeta (ζ). The component stars are both magnitude 4.5, and they lie 1.0 arcseconds apart. A good 5-inch telescope can split it. Zeta's orbital period is 123 years; by the year 2000 the stars will be 0.8 arcseconds apart.

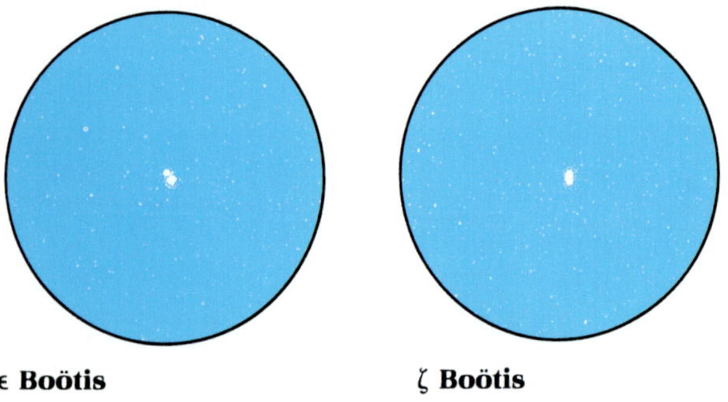

ϵ Boötis **ζ Boötis**

Corona Borealis, the Northern Crown

Corona is a pretty constellation, and very easy to recognize. Its most remarkable features are two variable stars, R and T. R is the prototype of its class, the R CrB stars. Normally it shines at maximum brightness—5.7 magnitude—but from time to time it suddenly fades out of sight, and recovers only after weeks or months. Keep a watch on it.

T is a recurrent nova. It nova'd in 1946, rising from eleventh to second magnitude, then faded just as it had 80 years earlier. Keep a watch on T's location—there's no telling when it will again brighten to naked-eye visibility.

Serpens Caput, the Head of the Snake (see star chart 15 for Serpens Cauda)

The glory of Serpens is M5, the second-brightest globular cluster in the northern sky. With a total brightness of sixth magnitude, M5 is visible with binoculars as a faint patch of light. Through a medium-to-large telescope, it's an impressive ball of stars. With small telescopes, you will just be able to "resolve" the cluster into stars at high magnification.

Beta (β) and delta (δ) are both doubles. Beta has a tenth-magnitude companion 30 arcseconds away; delta consists of fourth- and fifth-magnitude stars 4 arcseconds apart. Because it's in the head of the Serpent, the variable star R is easy to find. R is a Mira-type variable, and ranges from 5.2 at maximum to 14 magnitude at minimum. Check on R Serpentis whenever you observe R and T CrB.

The stars in star chart 11 pass overhead at:

February 21 at 4 A.M.; March 7 at 3 A.M.; March 23 at 2 A.M.; April 7 at 2 A.M.; April 22 at 1 A.M.; May 7 at midnight; May 23 at 11 P.M.; June 7 at 10 P.M.

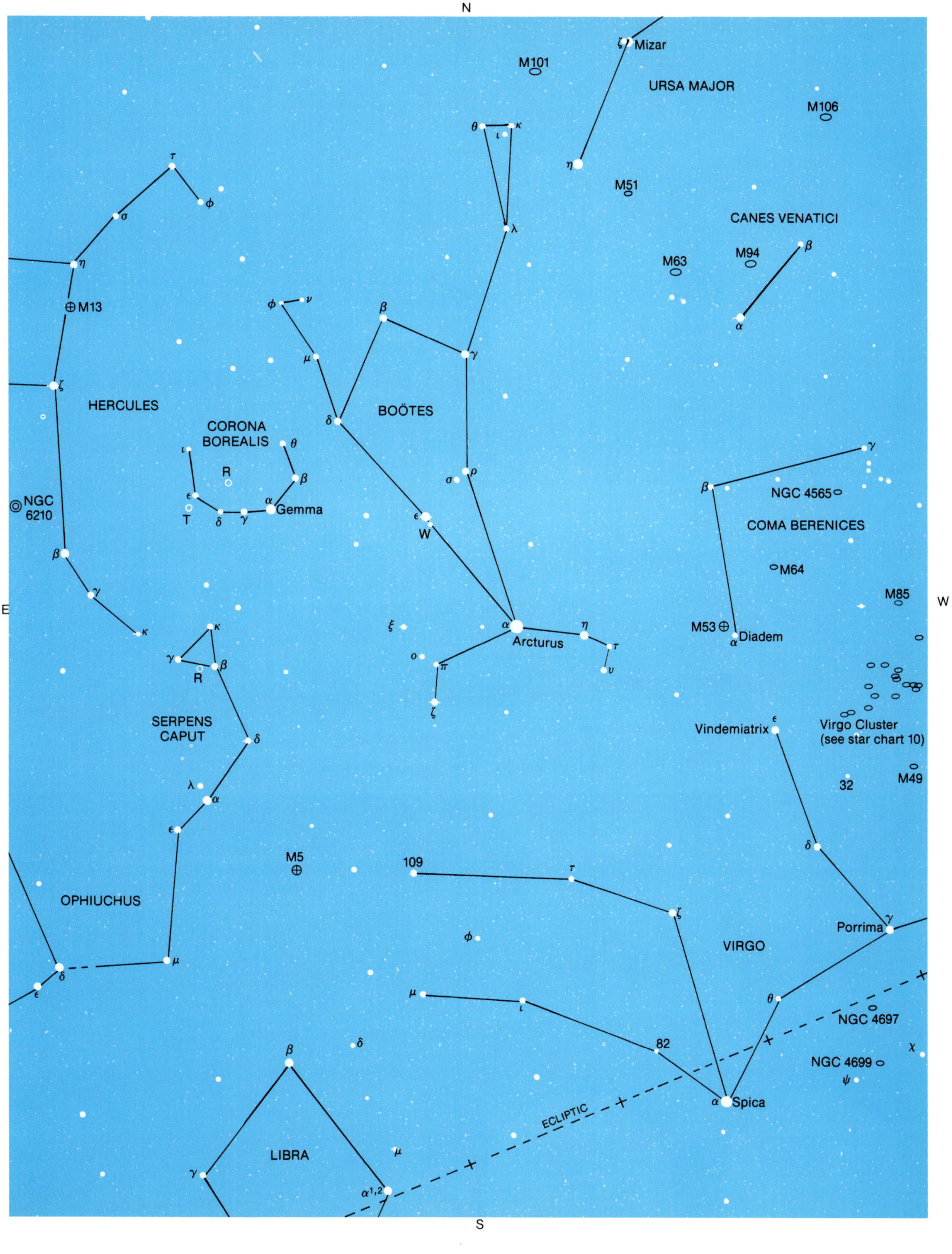

N
E
W
S
Mizar
M101
URSA MAJOR
M106
M51
CANES VENATICI
M63
M94
M13
HERCULES
CORONA
BOREALIS
BOÖTES
R
T
Gemma
NGC
6210
W
NGC 4565
COMA BERENICES
M64
M85
Arcturus
M53
Diadem
R
SERPENS
CAPUT
Vindemiatrix
Virgo Cluster
(see star chart 10)
32
M49
M5
109
OPHIUCHUS
Porrima
VIRGO
NGC 4697
82
NGC 4699
ECLIPTIC
Spica
LIBRA

STAR CHART 12
Facing south

Between the stars of spring and the stars of summer, south of the corn goddess Virgo, the last of the Water Snake slithers by. We see the stars of the ancient constellation, Claws of the Scorpion, which was renamed Libra, the Roman Scales of Justice, a mere 2,000 years ago. Low in the south, visible dimly through the horizon haze, we spy the Centaur and the Wolf he holds.

Libra, the Scales

Zubenelgenubi and Zubeneschamali are the Arabic names of alpha (α) and beta (β) Librae; these names mean northern claw and southern claw, respectively. Alpha is a wide double that you might be able to split naked-eye, and can certainly split easily with any binocular. The two stars shine at 2.8 and 5.2 magnitude and lie 231 arcseconds apart.

Iota has a wide unrelated naked-eye companion, a telescopic double, and a close binary star. The companion, sixth-magnitude 25 Librae, lies ¼° northeast of iota (ι), and is easily seen with a binocular. A ninth-magnitude companion lies 58 arcseconds from iota, fairly easy with a small telescope. Iota is also a binary, with an orbital period of 22 years, the stars separated by 0.1 arcseconds, well beyond the range of any amateur's telescope.

For a challenge, try mu (μ) Librae, with 1.8 arcseconds beween its sixth- and seventh-magnitude components. Delta (δ) reveals itself binary because its stars eclipse one another. Every 2 days, 7 hours, 51 minutes, the star dims by one magnitude. Delta is an Algol variable.

Perhaps the best deep-sky object in Libra is the globular cluster NGC 5897, a ninth-magnitude glow lying on a line between sigma (σ) and gamma (γ) just about opposite iota (ι).

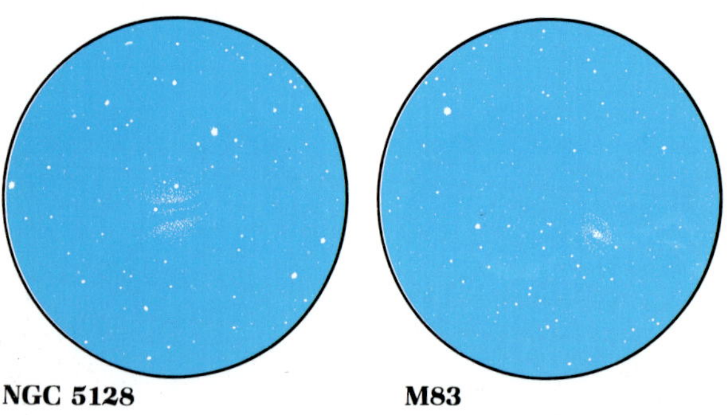

NGC 5128 **M83**

Centaurus, the Centaur

Centaurus is a Milky Way constellation just loaded with good things to see, but unfortunately it's too far south for most North American observers. Only from the Gulf States (Florida, Alabama, Mississippi, Louisiana, and Texas), New Mexico, Arizona, and southern California can its treasures be seen at all. From 45° north, the stars kappa (κ), nu (ν), phi (ϕ), and mu (μ) are just about lost in the horizon murk.

Centaurus's finest object is, without doubt, omega (ω) Centauri. Bright enough to have been given a star's name, this globular cluster makes the northern hemisphere's famed Hercules Cluster (M13) look sick by comparison. Through a large amateur telescope, omega is a glowing sphere of stars, a glorious sight.

The next best attraction is NGC 5128, a large, bright, elliptical galaxy with a peculiar dark lane of dust. With a medium-sized telescope, it's about half the diameter of the full moon, and the dust lane is clearly visible.

Star 3 is a double of two blue-white stars separated by some 8 arcseconds, located about as far north as anything in Centaurus.

The star nearest our sun is alpha (α) Centauri, four light-years distant. Unfortunately, it's below the horizon for most of North America.

Lupus, the Wolf

Lupus represents a wolf the Centaur has captured. If you'd like to see something in it, train your telescope on eta (η), a double. It has an eighth-magnitude companion 15 arcseconds away, but requires a clear and steady night to split because it's so low in the sky.

Hydra, the Water Snake (see also star charts 6 and 8)

The tail end of Hydra includes the beautiful M83, a large but rather dim face-on spiral galaxy; the Mira-type variable star R, which rises to third magnitude and falls to eleventh every 390 days; and the double 54, with fifth- and seventh-magnitude stars separated by 8 arcseconds.

The stars in star chart 12 pass overhead at:

February 21 at 5 A.M.; March 7 at 4 A.M.; March 23 at 3 A.M.; April 7 at 3 A.M.; April 22 at 2 A.M.; May 7 at 1 A.M.; May 23 at midnight; June 7 at 11 P.M.; June 22 at 10 P.M.

N
Vindemiatrix
SERPENS
CAPUT
OPHIUCHUS
M5
109
VIRGO
Porrima
NGC 4697
NGC 4699
82
Spica
ECLIPTIC
M104
Zubeneschamali
M107
48
LIBRA
Zubenelgenubi
25
NGC 5897
NGC
4361
CORVUS
R
W
E
M80
Dschubba
54
56
58
52
47
HYDRA
M83
M68
Antares
M4
3
1
2
SCORPIUS
CENTAURUS
d
LUPUS
NGC 5128
NGC 6231
NGC 5460
S

STAR CHART 13
Overhead, facing south

Before the summer Milky Way on late spring nights lies a region speckled with not-very-bright stars. Only zero-magnitude Vega, the luminary of the Lyre, sparkles brightly in this area. At first glance, the space seems empty, but two mighty heroes stride here. Overhead is Hercules, the kneeling hero, phantom-like in his ability to slip out of sight even as you're watching him. To the south of these groups is the barrel-chested serpent handler, Ophiuchus.

Hercules, the Hero

M13, the best known and brightest globular cluster in the northern skies, lies in Hercules. It is two-thirds of the way from zeta (ζ) to eta (η), on a line between them. You should be able to see it as a little round glow with a good binocular. If you're new to stargazing, you may find it difficult to reconcile a word like "bright" with the feeble glow you see in the sky. All I can say is: see it sometime with a big telescope under a dark sky. Then you'll understand.

Hercules offers a variety of doubles. Alpha (α), also known as Rasalgethi, is a close binary of contrastingly colored stars. The bright star is an orangy irregular variable star. Its companion, which looks blue or greenish by comparison, lies 4.6 arcseconds from it and circles it every 3,600 years. Delta (δ) is an optical double—its stars are unrelated. The star 95 Herculis is a fifth-magnitude pair matched in brightness but differing in color, and separated by 6.3 arcseconds. For a challenge, can you split zeta (ζ)? The primary is 2.9 and the secondary 5.5, but they're a mere 1.6 arcseconds apart. Zeta's orbital period is just 35 years, so look in on it every few years. Over the decades, you'll see it change.

Lyra, the Lyre

Lyra is another constellation that is, like Hercules, full of beautiful doubles, including the famous double-double, epsilon (ϵ) Lyrae. In addition, Lyra offers observers the Ring nebula, one of the largest and brightest planetary nebulae in the sky.

Let's start with epsilon (ϵ) Lyrae. It's called the double-double because it consists of a pair so wide that people with good eyes can see that it's double without optical aid. It's easily split with binoculars and opera glasses. With any reasonably powerful telescope, you'll be able to split each component of the wide double into another double. That's four stars in all. The main pair, epsilon-1 and epsilon-2, are separated by 207 arcseconds. These are split into pairs 2.3 and 2.6 arcseconds wide, respectively.

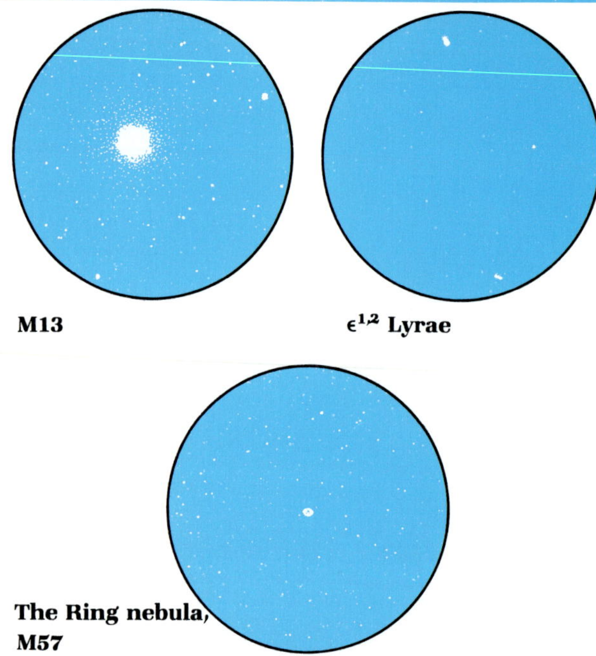

M13

$\epsilon^{1,2}$ Lyrae

The Ring nebula, M57

Beta (β) is a wide, easy double with a bluish eighth-magnitude companion, and the primary is an eclipsing variable star. It cycles from 3.4 to 4.3 magnitude in a little under 13 days. Why not watch it every clear night for a month or two? Compare its brightness with that of zeta (ζ). By the way, zeta is yet another wide double!

Lyra's *pièce de résistance* is the Ring nebula, M57. Tucked midway between gamma (γ) and beta (β) in the bottom of the lyre, this little doughnut of light is a shell of expanding gas puffed off by an ancient and dying star. You might miss it the first few times you look. It's smaller (only 70 by 150 arcseconds) and dimmer (ninth magnitude) than you'd expect for so famous an object. After locating it at low power, examine it carefully at high magnification. It really is a little ring.

While you're at it, don't miss the globular cluster M56.

The stars in star chart 13 pass overhead at:

April 7 at 5 A.M.; April 22 at 4 A.M.; May 7 at 3 A.M.; May 23 at 2 A.M.; June 7 at 1 A.M.; June 22 at midnight; July 7 at 11 P.M.; July 23 at 10 P.M.; August 7 at 9 P.M.

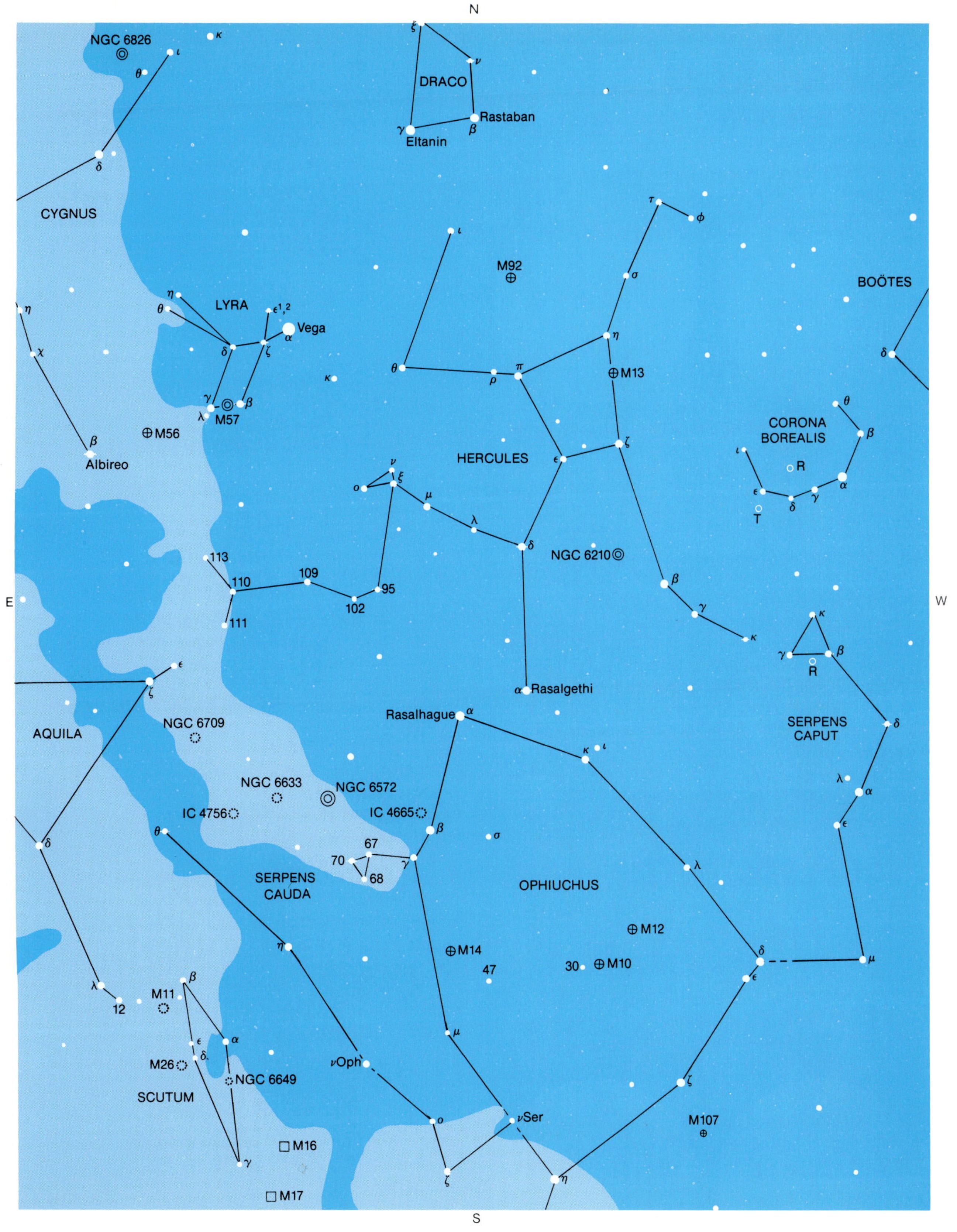

N
E
W
S
DRACO
Rastaban
Eltanin
NGC 6826
CYGNUS
LYRA
Vega
M57
M56
Albireo
M92
M13
HERCULES
BOÖTES
CORONA
BOREALIS
NGC 6210
113
110
109
102
95
111
Rasalgethi
Rasalhague
AQUILA
NGC 6709
NGC 6633
NGC 6572
IC 4756
IC 4665
SERPENS
CAPUT
67
70
68
SERPENS
CAUDA
OPHIUCHUS
M12
M14
M10
47
30
12
M11
M26
NGC 6649
SCUTUM
νOph
νSer
M107
M16
M17

STAR CHART 14
Facing south

Few areas of the sky are as dramatic as this brightest segment of the Milky Way. The outline of the Serpent Handler, Ophiuchus, marks a large area of sky with relatively few bright stars. Below the giant, orange Antares shines in the heart of the Scorpion. A great curve of stars marks the chitinous tail drooping toward the horizon, then curving up, stinger held at ready.

Scorpius, the Scorpion

Scorpius is one of the most magnificent areas in the entire sky, in the direction of the center of the galaxy. It is filled with bright Milky Way star clouds, clouds of absorbing dust, and sparkling star clusters.

Begin exploring Scorpius with Antares, a cool red giant star whose name means "the rival of Mars" for its deep orange hue. Antares is variable, ranging from 0.8 to 1.8 magnitude, somewhat irregularly, with a period of 5 years. Antares is also a binary. Its hot, blue companion orbits it in 878 years, shining at magnitude 5.4, and currently lies 2.6 arcseconds from its primary. Because of the dazzle from Antares, the companion is rather hard to spot unless the air is exceptionally steady.

Near the heart of the Scorpion are two globular clusters, M4 and M80. M4 is by far the brighter and larger of the two, and quite close—a mere 6,800 light-years away. It can be seen with binoculars; small telescopes resolve it into stars. M80 is four times more distant and correspondingly fainter.

Behind the Scorpion's tail you'll find two huge, bright open star clusters, M6 and M7. Both are naked-eye objects, and impressive with binoculars. The brightest of their hundred or so stars shine at sixth magnitude. M7 is the larger and brighter of the two. Enveloping the wide double star zeta (ζ) in the tail is the cluster NGC 6231, which is just part of a larger association of hot young stars.

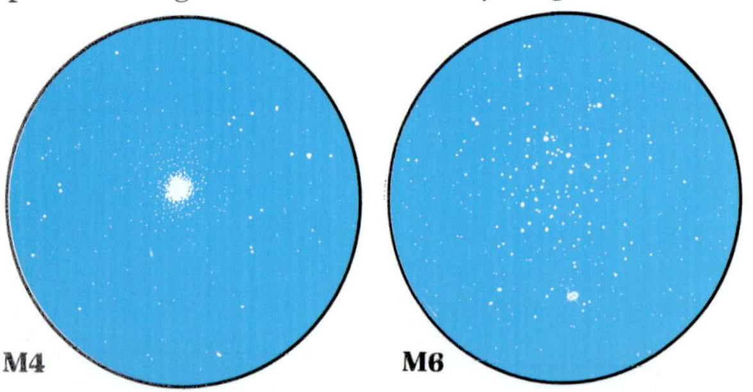

M4 M6

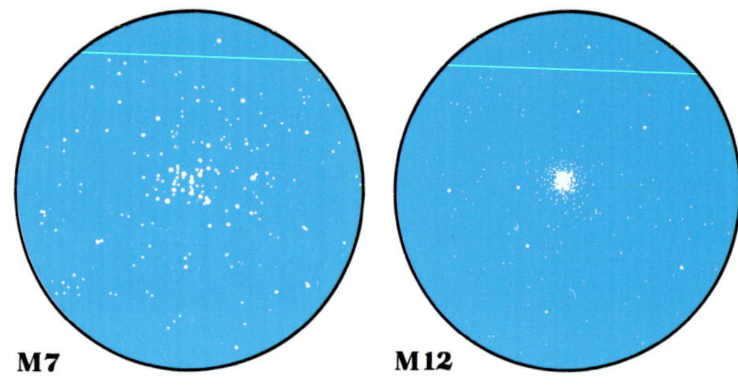

M7 M12

Ophiuchus, the Serpent Handler

Globular clusters orbit the center of our Milky Way galaxy. When we look toward the center, provided dark dust clouds don't block the view, we can see them out there. And that's the case with Ophiuchus. Using the chart, take yourself on a tour of the Ophiuchus globular clusters: M9, M10, M12, M14, M19, M62, and M107. (And don't forget there are lots of globulars in Scorpius and Sagittarius, too.) All seven have roughly the same brightness—seventh magnitude—and lie between 14,000 and 35,000 light-years away.

The center of our galaxy lies 28,000 light-years away, in a direction that lies between the stars 45 Ophiuchi and gamma (γ) Sagittarii, marked by a + symbol on the star chart. You can't see anything there because intervening dust clouds absorb all the light from our galaxy's center, but astronomers observe radio waves coming from it.

In a little spur of Milky Way you'll find two open clusters, NGC 6633 and IC 4665. IC 4665 is a large bright cluster right next to beta (β) Ophiuchi. It's easy to see with binoculars. NGC 6633 is smaller and dimmer, but still pretty easy to locate.

While you're in Ophiuchus, don't miss the doubles 36 and 70. Star 36 consists of two identical fifth-magnitude orange stars 5 arcseconds apart, while 70's are fourth- and sixth-magnitude, and much closer. With an orbital period of 88 years, here's one you can watch change. Between 1990 and 1995, the separation will increase from 1.5 to 2.5 arcseconds. Star 70 is one of the sun's neighbors; it lies only 16.8 light-years away in space.

The stars in star chart 14 are highest at:

April 7 at 5 A.M.; April 22 at 4 A.M.; May 7 at 3 A.M.; May 23 at 2 A.M.; June 7 at 1 A.M.; June 22 at midnight; July 7 at 11 P.M.; July 23 at 10 P.M.; August 7 at 9 P.M.

N
S
E
W
SERPENS CAPUT
OPHIUCHUS
SERPENS CAUDA
SCUTUM
LIBRA
SCORPIUS
SAGITTARIUS
CORONA AUSTRALIS
LUPUS
CENTAURUS
NGC 6633
IC 4756
NGC 6572
IC 4665
M5
M12
M14
M10
M11
M26
NGC 6649
ν Oph
ν Ser
M107
M16
M17
M18
M24
M23
M25
M9
ECLIPTIC
M21
M20
M8
M22
M28
M80
Dschubba
NGC 5897
Antares
M4
M19
Nunki
X
Galactic Center
M62
M6
M54
M70
M69
M7
Shaula
G
Q
NGC 6231
NGC 6541

STAR CHART 15
Facing south

Chasing Scorpius on warm summer nights is Sagittarius, the Archer, bow taut, ready to let fly. Like a mighty river, the Milky Way flows from the zenith through Aquila, Scutum, Sagittarius, over the tail of Scorpius, and into the southern horizon.

Sagittarius, the Archer

No other region of the sky is so filled with magnificent objects for the naked eye, binocular, and telescope as the Milky Way in Sagittarius and Scorpius. On a clear, dark, moonless night, sweep this region with low power and marvel at the glittering fields that pass before your eyes.

The starclouds of the Milky Way are most apparent with no optical aid at all. The Great Sagittarius Starcloud lies just east of gamma (γ) like celestial steam coming from the spout of the Sagittarius "teapot" asterism. It's about 7° long and 4° wide. A little farther north, enveloping the open cluster M24, is the Small Sagittarius Starcloud, roughly 2° by 3°. (A third major starcloud, the Scutum Starcloud, underlies the constellation Scutum.)

The Lagoon nebula, M8, is a faint patch to the naked eye, a twinkle of stars enmeshed in a glow through binoculars, and a magnificent star cluster floating in a glowing cloud with a telescope. This great mass of gas gave birth to the star cluster in it. A little north of the Lagoon is the Trifid nebula, M20, split into three by overlying lanes of dark material. The Trifid is much fainter than the Lagoon. Farther north, you'll find the bright, checkmark-shaped Omega nebula (M17) as a glow in your telescope's finder, and a curdled, glowing mass in the eyepiece of the telescope.

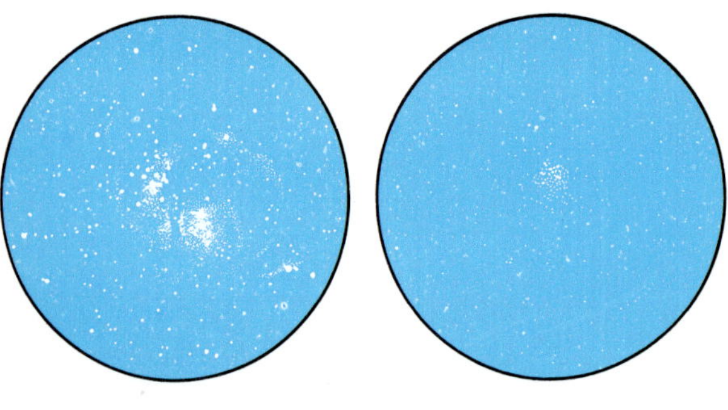

The Lagoon nebula, M8 **M11**

The Milky Way is studded with open clusters. The brightest is M25, a cluster of a few dozen seventh- and eighth-magnitude stars. By contrast, M23 contains hundreds of ninth- and tenth-magnitude stars. You'll want to search out M18, M21, and inconspicuous M24, nearly lost among the stars in the Small Sagittarius Starcloud, too.

M22, just north and east of lambda (λ), is the brightest of Sagittarius's many globular clusters, at fifth magnitude, and one of the best and brightest in the whole sky. M28 and M55 are seventh-magnitude globulars, while M54, M69, and M70 shine at eighth magnitude. You can see them all with binoculars.

Inveterate double-star observers will want to split eta's (η) third- and eighth-magnitude components; they lie 3.6 arcseconds apart. Eta is a deep orange color and its faint companion is blue. Near the spout of the "teapot" are two bright Cepheid variable stars, W and X Sagittarii. Both peak at 4.3 magnitude, and dip to fifth magnitude, with periods just over seven days.

Scutum, the Shield

In the Scutum Starcloud you'll find a fine open cluster, M11, called the Wild Duck Cluster because its swept-back shape resembles a group of migrating fowl. Once you've located it, switch to high magnification and enjoy the sight of its hundreds of faint stars. M26 and NGC 6649 are dimmer, poorer clusters nearby.

Serpens Cauda, the Serpent's Tail (see star chart 11 for Serpens Caput)

The Serpent's Tail runs up the Great Rift in the Milky Way, and so contains relatively little of interest. M16 is the Eagle nebula, a faint nebula surrounding a star cluster. It lies close to M17. IC 4756 is a bright cluster for binoculars, larger than the lunar disk and quite bright. Theta (θ), the star at the tip of the Tail, is a double, its two white stars separated by 22 arcseconds.

The stars in star chart 15 are highest at:

April 7 at 4:30 A.M.; April 22 at 5:30 A.M.; May 7 at 3:30 A.M.; May 23 at 2:30 A.M.; June 7 at 1:30 A.M.; June 22 at 12:30 A.M.; July 7 at 11:30 P.M.; July 23 at 10:30 P.M.; August 7 at 9:30 P.M.

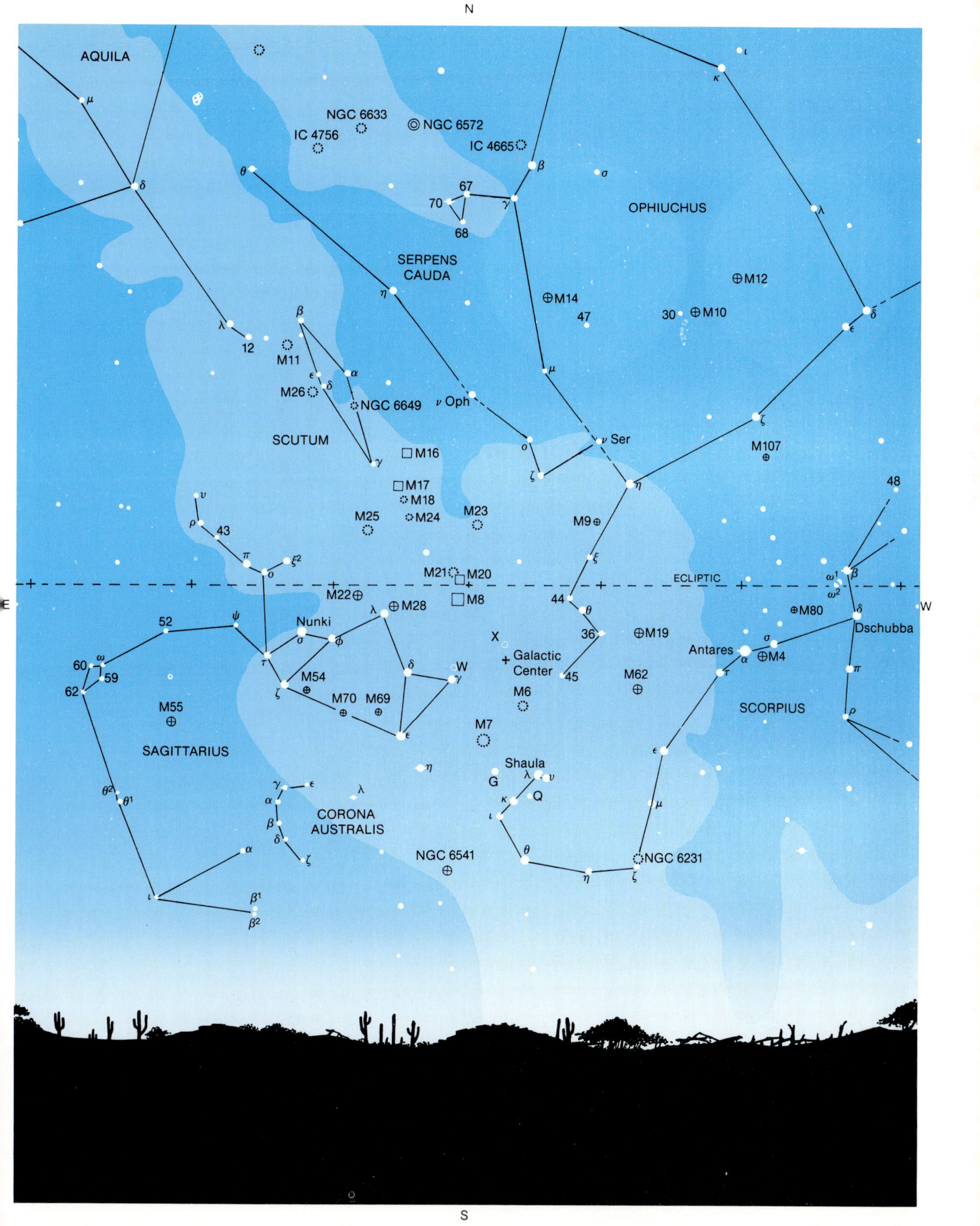
N
S
E
W
AQUILA
NGC 6633
IC 4756
NGC 6572
IC 4665
67
70
68
OPHIUCHUS
SERPENS
CAUDA
M12
M14
47
30
M10
12
M11
M26
NGC 6649
ν Oph
SCUTUM
ν Ser
M16
M107
M17
M18
48
M25
M24
M23
M9
43
ECLIPTIC
M21
M20
M22
M28
M8
44
M80
Dschubba
52
Nunki
36
M19
Antares
M4
60
59
X
Galactic
Center
62
M54
45
M62
M70
M69
M6
M55
SCORPIUS
M7
SAGITTARIUS
Shaula
G
Q
CORONA
AUSTRALIS
NGC 6541
NGC 6231

STAR CHART 16
Facing north

On midsummer nights, the dragon, Draco, and the house-shaped king, Cepheus, reach their highest. These ancient constellations sometimes seem to trick the eye, suddenly dissolving, then reappearing as the observer's eye and brain once again find the turns and twists of the dragon's body or the geometric shape of Cepheus.

Cepheus, the King

Cepheus is full of interesting stars, particularly three variable stars that serve as prototypes of their class. These are delta (δ) Cephei, the prototype Cepheid variable; beta (β) Cephei, a Cepheidlike variable with a small range of variation; and mu (μ) Cephei, a supergiant star and semiregular red variable.

Cepheid variables are rather old stars that, for a few million years late in their lives, swell and collapse with metronomic regularity. Their brightness changes in a regular pattern. Delta (δ) was the first such star found. You can easily track this star's cycle in a little over 5⅓ days. Here's the trick: the nearby stars zeta (ζ) (3.35 magnitude) and epsilon (ϵ) (4.19) are just about as bright and faint as delta gets. If you call the brightness of zeta 1 and that of epsilon 10, you can estimate delta's brightness on a scale of 1 to 10. Try it. Record your estimates in your observing notebook. After you've collected a few months' data, graph delta's variation.

William Herschel called mu (μ) Cephei the Garnet Star, and if you look at it with a telescope, you'll see that it's a deep orangy-red color. Mu is a supergiant star, and it slowly and somewhat irregularly varies, from 3.5 to 5.1 magnitude and back, in about two years. Mu is just about the reddest star in the sky. Nearby beta (β) Cephei is about as blue as a star can be. Look at both; compare them.

Beta is a double as well as a variable. Its eighth-magnitude companion lies 13 arcseconds from the

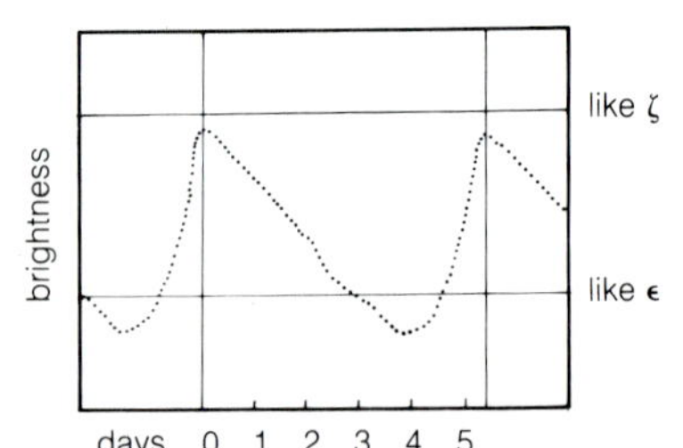

δ Cephei's variation

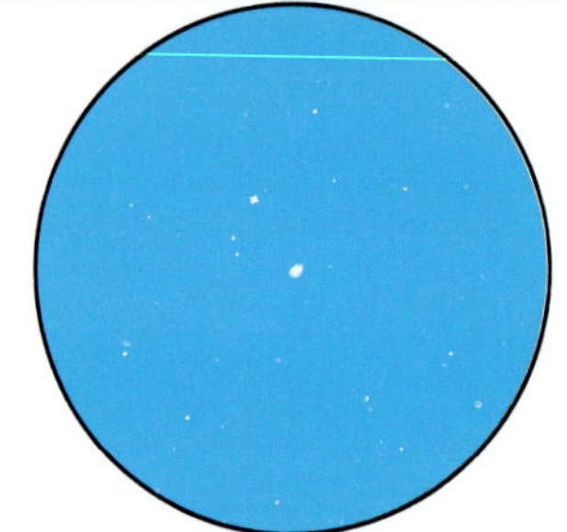

NGC 6543

third-magnitude variable-star primary. Kappa (κ) is quite similar in brightness, but a little closer. The magnitudes of the stars are 4.4 and 8.4, and they lie 7 arcseconds apart. Xi (ξ) presents a nice color contrast. The primary is a fourth-magnitude blue star, and the secondary pale yellow. Their separation is 8 arcseconds.

Draco, the Dragon

From its head trapped under the foot of Hercules, gangly Draco winds around Ursa Minor. Although the head consists of four stars, if you ignore dim nu (ν), the remaining three form a striking lozenge with neighboring iota (ι) Herculis.

Because our planet's axis of rotation slowly precesses like a gigantic toy top, the polestar changes. The precessional cycle takes about 26,000 years. Thuban (alpha (α) Draconis) was the polestar 5,000 years ago.

With a binocular, 17 is a wide double of fifth-magnitude blue-white stars. Their angular separation is 90 arcseconds. With a telescope, you can split 17 again, this time into two stars separated by 3 arcseconds. Star 39 is another triple. The primary is fifth magnitude. One companion is an eighth-magnitude star 4 arcseconds away; the other a seventh magnitude star 89 arcseconds distant. Psi (ψ) and nu (ν) are both moderately wide doubles. Psi's components are magnitudes 4.9 and 6.1 with a separation of 30 arcseconds; nu's are both 4.9, and separated by 62 arcseconds.

Be sure to look for NGC 6543, a striking planetary nebula. Through a telescope, it looks like a ninth-magnitude blue-green planet 18 arcseconds in diameter, but it has nothing to do with planets at all. This puff of material is the outer envelope of a dying star, ejected in the final throes of its evolution.

The stars in star chart 16 are highest at:

May 23 at 4 A.M.; June 7 at 3 A.M.; June 22 at 2 A.M.; July 7 at 1 A.M.; July 23 at midnight; August 7 at 11 P.M.; August 23 at 10 P.M.; September 7 at 9 P.M.; September 22 at 8 P.M.

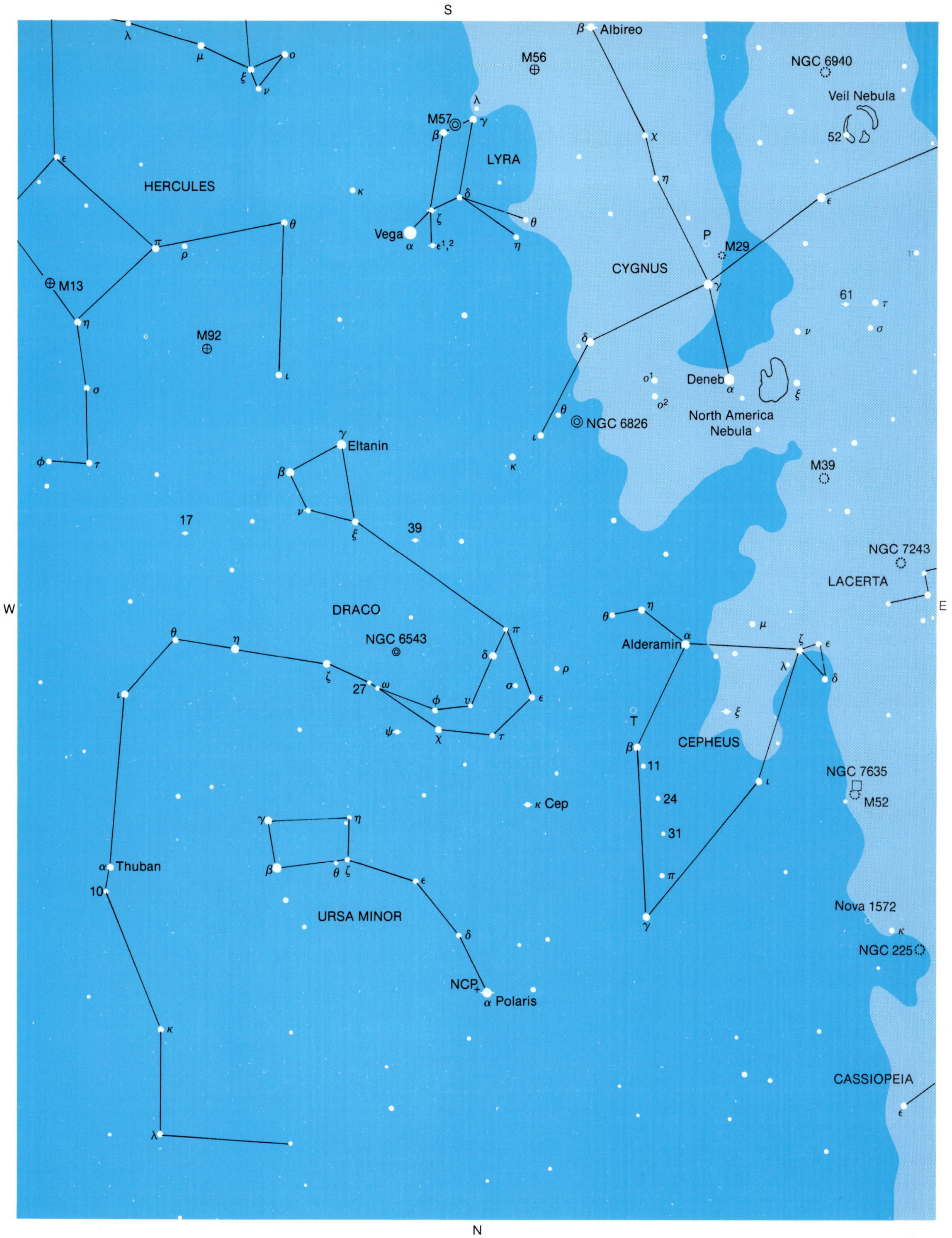

S

W

E

N

HERCULES

M13

M92

LYRA

Vega

M57

M56

CYGNUS

Albireo

Deneb

North America Nebula

NGC 6826

M29

P

NGC 6940

Veil Nebula

52

61

M39

Eltanin

17

39

DRACO

NGC 6543

27

Alderamin

CEPHEUS

T

11

24

31

κ Cep

LACERTA

NGC 7243

NGC 7635

M52

Nova 1572

NGC 225

CASSIOPEIA

Thuban

10

URSA MINOR

NCP

α Polaris

STAR CHART 17
Overhead, facing south

On midsummer nights let your thoughts go. Against the glowing stream of the Milky Way imagine the Swan and Eagle winging south. Can you find the stars that comprise the Dolphin, Arrow, Foal, and Little Fox?

Cygnus, the Swan

For the beginner, Cygnus offers a superb double star: Albireo, or beta (β) Cygni. Binoculars will split this gold and blue pair consisting of third- and fifth-magnitude stars 34 arcseconds apart. Albireo is magnificent through any telescope. While you're looking, observe 61, too. This double, only 11 light-years from us in space, consists of a fifth-magnitude star and a sixth-magnitude companion 30 arcseconds apart in the eyepiece. Both are gold in color.

The Milky Way in Cygnus and Aquila is magnificent. It flows in two wide channels the length of both constellations. The Cygnus Starcloud is the branch between gamma (γ) and beta (β) Cygni, certainly the richest area of the northern Milky Way. It is sheer pleasure to scan with large binoculars and low-power telescopes. The Great Rift, which divides the two starry streams, opens near Deneb, continues the length of Cygnus, through Vulpecula, and all the way down to Serpens Cauda.

Two ripples in the Milky Way stream bear Messier designations: M29 and M39. M29 is a few dozen stars just south of gamma (γ) Cygni, on the boundary of the Great Rift, a bit of glitter in a region filled with sparkling stars. M39 consists of fifty stars scattered over an area the size of the full moon.

In addition to stars, the Milky Way glows with clouds of luminous hydrogen gas called *emission nebulae*. The best known, the North America nebula, a glowing patch visible with binoculars, lies between Deneb and xi (ξ) Cygni at the head of the Great Rift.

Near 52 and east of it is the Veil nebula, a supernova remnant. This 10,000-year-old shell of gas was ejected in the annihilation of a massive star that had consumed its nuclear fuel. Visually, the Veil consists of two glowing arcs and a triangular patch. It is a difficult object—large, diffuse, and faint.

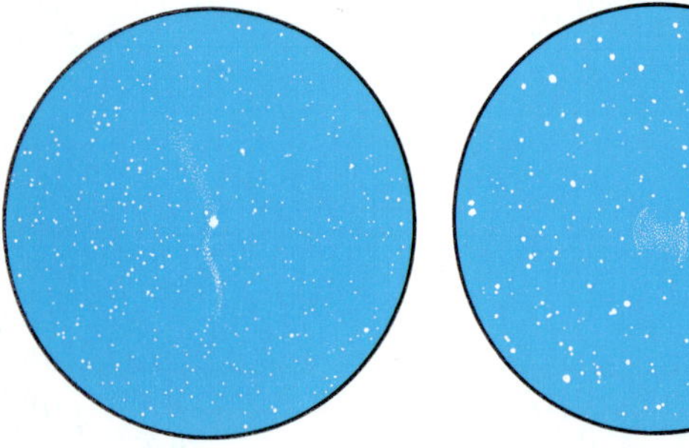

The Veil nebula

The Dumbbell nebula, M27

Vulpecula, the Little Fox

The Dumbbell nebula, M27, is little Vulpecula's prime attraction. This exceptionally large and bright planetary nebula shines with the light of a seventh-magnitude star, and is 350 arcseconds in diameter. It lies just north of gamma (γ) Sagittae.

Delphinus, the Dolphin

The star in the Dolphin's nose, gamma (γ), is a pretty double: 4.5- and 5.5-magnitude stars separated by 10 arcseconds. The primary is yellow-tinged, and the secondary is white.

Sagitta, the Arrow

Right in the shaft of the Arrow you'll find an eighth-magnitude globular cluster, the seventy-first object in Messier's famous catalog of diffuse objects. M71 lies 13,000 light-years from us.

Equuleus, the Foal

Epsilon (ε) is a challenging double star, two sixth-magnitude stars separated by 1 arcsecond. Its orbital period of 101 years means that you can see the distance and angle between the stars change if you observe it for a few decades.

Aquila, the Eagle

Beta (β) Aquilae is a challenging double star. The primary is magnitude 3.7, but its companion is eleventh magnitude. They lie 13 arcseconds apart. Can you see the companion? Here's a project to try: Eta (η) is a Cepheid variable star, ranging from 3.4 to 4.3 magnitude in a period of seven days. Compare its brightness nightly with beta (β) (3.7), iota (ι) (4.4), and theta (θ) (4.4) for a few months; then plot the curve of eta's brightness against time.

The stars in star chart 17 pass overhead at:

June 7 at 4 A.M.; June 22 at 3 A.M.; July 7 at 2 A.M.; July 23 at 1 A.M.; August 7 at midnight; August 23 at 11 P.M.; September 7 at 10 P.M.; September 22 at 9 P.M.; October 7 at 8 P.M.

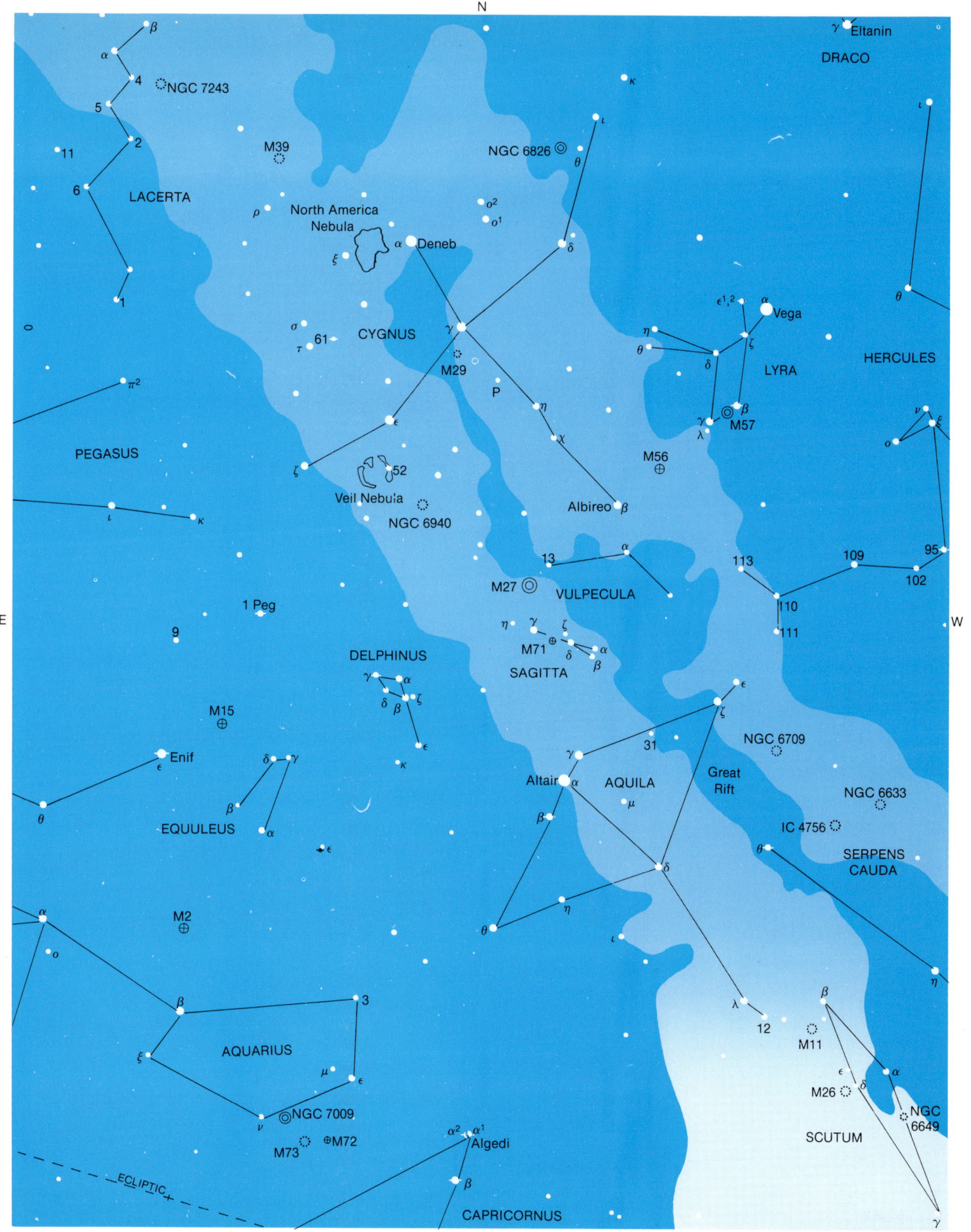
N
S
E
W
DRACO
Eltanin
NGC 7243
M39
NGC 6826
LACERTA
North America
Nebula
Deneb
CYGNUS
Vega
LYRA
HERCULES
M29
P
M57
PEGASUS
M56
Veil Nebula
NGC 6940
Albireo
M27
VULPECULA
1 Peg
DELPHINUS
M71
SAGITTA
M15
NGC 6709
Enif
Altair
AQUILA
Great
Rift
NGC 6633
EQUULEUS
IC 4756
SERPENS
CAUDA
M2
M11
AQUARIUS
M26
NGC
6649
NGC 7009
SCUTUM
M73
M72
Algedi
ECLIPTIC
CAPRICORNUS

STAR CHART 18
Facing south

Following the bright stars of the Archer and soft glow of the Milky Way is a lusterless section of sky, prelude to the dim "watery" constellations of autumn. Below the "teapot" of Sagittarius, you'll find a delicate arc of stars marking Corona Australis, the Southern Crown. To the left of Sagittarius is the broad Cheshire cat grin of Capricornus, its angular outline as suggestively goatlike today as it was 4,000 years ago.

Capricornus, the Sea Goat

Capricornus has lots of doubles, starting with alpha (α), a naked-eye pair. The eastern star is alpha-1, and the western star, alpha-2. Alpha-2 is about twice as bright. They lie one-tenth of a degree apart. Both stars are themselves doubles.

Alpha-1's companion is a ninth-magnitude star some 45 arcseconds distant, an easy job for small telescopes. Alpha-2 has an eleventh-magnitude companion only 6 arcseconds from it, and is quite difficult to split. That companion star is itself double, the two stars separated by a mere 1.2 arcseconds. It's beyond most amateur telescopes.

Beta (β) is a wide binocular double: 205 arcseconds. The companion star is sixth magnitude, and very blue compared to its primary's yellowy tinge. The two stars look quite splendid through a small telescope at low magnification.

Each of the stars in the little triangle of pi (π), rho (ρ), and omicron (o) is multiple. Pi's 8.9-magnitude companion lies 3.2 arcseconds from it, a fairly difficult double. Rho is a triple, with a very close tenth-magnitude secondary (no hope!) and a wide seventh-magnitude tertiary some 4 arcminutes away. Omicron is an easy one, the two stars separated by 22 arcseconds and about the same brightness.

South of delta (δ) and right beside 41 Capricorni is M30, a moderately bright globular cluster. Compare it with M2 and M72 in Aquarius. Which is best?

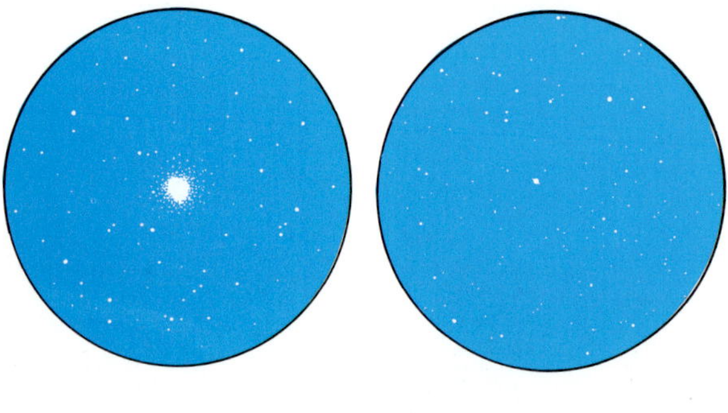

M2 NGC 7009, the Saturn nebula

Corona Australis, the Southern Crown

South of Sagittarius you'll find the Southern Crown, alongside the starclouds of the Milky Way. It's definitely a constellation for observers in the southern states, but northerners may glimpse it on especially clear nights. Its best deep-sky object is the sixth-magnitude globular cluster NGC 6541.

Aquarius, the Water Man (see also star chart 20)

Eastern Aquarius offers several excellent telescopic objects, one nonobject, and an interesting comparison of rich and poor globular clusters. With a binocular or small telescope, you'll see a faint glow 5° north of the star beta (β)—the globular cluster M2. With the total light of a 6.5-magnitude star spread over an 8-arcminute disk, the object is an easy one. It lies approximately 37,000 light-years away. After a careful look at M2, locate M72 some 3° south of epsilon (ϵ) and compare them.

While you're at it, try to find M73 1° east of M72. Do you see anything? Charles Messier thought he did—but all a careful search reveals is a little clump of four tenth-magnitude stars. With his small telescope, Messier apparently mistook them for a nebula and added them to his list.

While you're in the vicinity, center on nu (ν) Aquarii; then look 1½° west for the Saturn nebula, NGC 7009 a bright planetary nebula that somewhat resembles a very dim image of Saturn. It looks rather green through a medium-sized telescope. The brightest part of the nebula is 25 arcseconds across, similar in angular size to Saturn.

Microscopium, the Microscope

South of Capricornus is an obscure little constellation honoring the microscope. A tenth-magnitude companion star lies 20 arcseconds from fifth-magnitude alpha (α).

The stars in star chart 18 are highest at:

June 7 at 4 A.M.; June 22 at 3 A.M.; July 7 at 2 A.M.; July 23 at 1 A.M.; August 7 at midnight; August 23 at 11 P.M.; September 7 at 10 P.M.; September 22 at 9 P.M.; October 7 at 8 P.M.

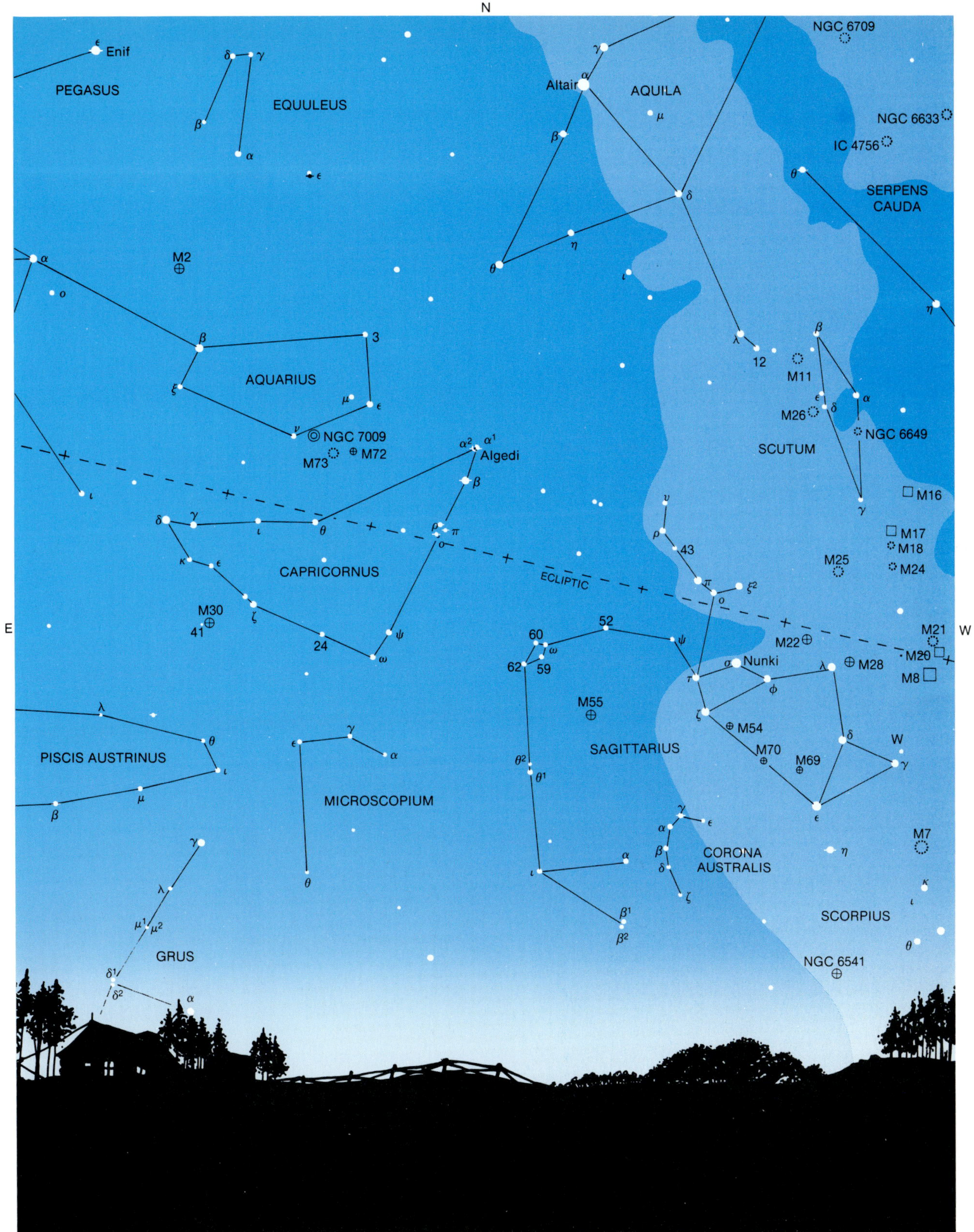

N
S
E
W
Enif
PEGASUS
EQUULEUS
Altair
AQUILA
NGC 6709
NGC 6633
IC 4756
SERPENS CAUDA
M2
AQUARIUS
3
12
M11
M26
SCUTUM
NGC 6649
NGC 7009
M73
M72
Algedi
M16
M17
M18
M24
M25
43
CAPRICORNUS
ECLIPTIC
M30
41
24
60
59
62
52
M22
M21
M20
M8
Nunki
M28
M55
M54
M70
M69
SAGITTARIUS
PISCIS AUSTRINUS
MICROSCOPIUM
CORONA AUSTRALIS
M7
SCORPIUS
GRUS
NGC 6541

STAR CHART 19
Overhead, facing south

The region around the Great Square of Pegasus is the most memorable part of the autumn sky. From one side of the square extend the head and legs of the horse itself; from the other side extend two graceful arcs of stars forming Andromeda. South of the Square is the Circlet of Pisces, the western fish of the Fishes constellation.

Pegasus, the Flying Horse

Visually, the Great Square is Pegasus's greatest asset. It encloses a region 13° high by 15° wide that contains few bright stars. Why not count those you can see? Don't be surprised if you can see only two from your suburban backyard. Under a dark country sky, you might see three dozen.

A bright and famous globular cluster, a pretty good spiral galaxy, a wide binocular double—that's what the constellation Pegasus offers the autumn stargazer. The star Enif, epsilon (ϵ) Pegasi, is the binocular double, the eighth-magnitude companion visible 142 arcseconds away from the yellow-orange primary star. While you're looking at Enif, shift your gaze 3° north and west to the globular cluster M15 (locate M15, near Enif, on star chart 17). With a total brightness equivalent to that of a sixth-magnitude star, M15 is clearly visible with binoculars.

NGC 7331 is smaller and fainter than its famous neighbor in Andromeda, but is more or less the same type of galaxy. If you have a big telescope, look for the group of faint galaxies called Stefan's Quintet. It lies ½° due south of NGC 7331. Warning: they're very faint.

Andromeda, the Chained Lady

The Andromeda galaxy is the prime attraction here. From alpha (α) skip to delta (δ) then beta (β), turn the corner, go by mu (μ) and a little past nu (ν). There M31 is visible to the naked eye as a faint glow. With binoculars, it is distinctly oval in shape. In size, it is roughly 3° long by 1° wide. Generally, beginners find this galaxy—a galaxy a lot like our own Milky Way—rather disappointing. They expect to see something big and bright. Yet with the exception of the Magellenic Clouds, our galaxy's nearby satellites, M31 is *the* biggest and brightest galaxy in the sky, 2 million light-years distant, and the only galaxy readily visible to the naked eye.

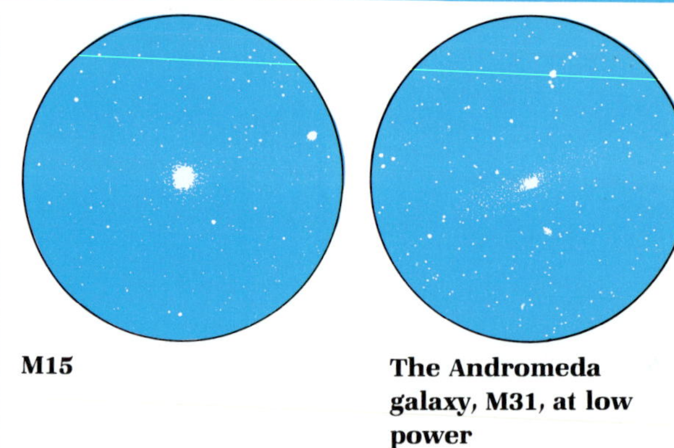

M15

The Andromeda galaxy, M31, at low power

Seeing it requires a clear, dark sky, low magnification, and a mind willing to accept that this dim glow in the sky is a galaxy like our own.

M31 has two small elliptical companion galaxies, M32 and NGC 205. Look for them, with a telescope, ½° on either side of M31.

Andromeda has some other goodies to offer. Gamma (γ) is a superb triple star. Ten arcseconds from the second-magnitude primary is a very close double, two sixth-magnitude stars only 0.5 arcseconds apart. Few amateur telescopes can split a star so close.

Planetary nebula NGC 7662 lies between omicron (ο) and iota (ι), a little closer to iota. Find its tiny disk (20 arcseconds across) at low power, then switch to high magnification for a better view. The galaxy NGC 891 (see star chart 22) is an edge-on spiral with a lane of dust down the middle.

Lacerta, the Lizard

Zigzagging little Lacerta does not have much to offer observers but one open star cluster, NGC 7243, with a few dozen scattered stars. The Polish astronomer Hevelius made up this constellation to fill the gap between Cygnus and Andromeda. Few astronomers even know where it is.

Pisces, the Fishes (see also star chart 22.)

Western Pisces has two double stars. Zeta (ζ) has blue and gold components 23 arcseconds apart, of magnitude 5.6 and 6.5 respectively. You'll find a ninth-magnitude companion to 34 Piscium just 8 arcseconds from the fifth-magnitude primary.

The stars in star chart 19 are overhead at:

July 23 at 4:30 A.M.; August 7 at 3:30 A.M.; August 23 at 2:30 A.M.; September 7 at 1:30 A.M.; September 22 at 12:30 A.M.; October 7 at 11:30 P.M.; October 23 at 10:30 P.M.; November 7 at 8:30 P.M.; November 22 at 7:30 P.M.

N
CASSIOPEIA
NGC 7789
M39
NGC 7243
R
LACERTA
ANDROMEDA
NGC 205
M31
M32
NGC 7662
TRIANGULUM
Mirach
M33
NGC 7331
Sirrah
PEGASUS
E
W
M74
PISCES
Algenib
Markab
Enif
Circlet
Water Jar
Vernal Equinox
ECLIPTIC
AQUARIUS
CETUS
S

STAR CHART 20
Facing south

This region lies south of Pegasus, far from the rich fields of the Milky Way. For observers in the northern states, the bright star Fomalhaut never rises high, and Sculptor, Phoenix, and Grus can't be seen. Nonetheless, there are treasures buried among these stars, especially when the sky is dark and transparent.

Aquarius, the Water Man (see also star chart 18)

The star at the center of the Water Jar, zeta (ζ), is a binary with an orbital period of 850 years. The pair is currently widening slowly; in 1990, their separation will be 2.0 seconds of arc. The components are closely matched in brightness—4.3 and 4.5—making this star a perfect test star for small refractors.

Aquarius contains a slightly abnormal long-period variable worth keeping under watch. Occasionally brightening to naked-eye visibility, and fading below the reach of binoculars, R is a symbiotic binary system from which matter is being ejected. Japanese astronomers saw the star flare in the year 930, so keep an eye on this one!

Tucked between Aquarius and Piscis Austrinus is the planetary nebula NGC 7293, the Helix nebula. Although its total light equals that of a magnitude-6.5 star, it is spread over an area half the diameter of the full moon! Binoculars show it as an exceedingly faint glow; through telescopes, the Helix looks its best at very low magnification. Starhop to it along the little chain of stars extending from delta (δ).

Piscis Austrinus, the Southern Fish

This vaguely fish-shaped constellation is dominated by first-magnitude Fomalhaut, the Fish's mouth. Only 22 light-years away, Fomalhaut is a rather ordinary blue-white star much like nearby Sirius.

For the telescope, beta (β) is an optical double star—that is, an unrelated pair that happen to line up. The brighter star is magnitude 4.4, the dimmer 7.9, separated by 30 seconds of arc. See if you can split this one with binoculars.

Gamma (γ) and eta (η) are considerably more challenging binaries. Gamma's magnitudes are almost exactly the same brightness as beta's, but the separation is only 4.2 seconds of arc. If the seeing is good, though, it will yield to small telescopes.

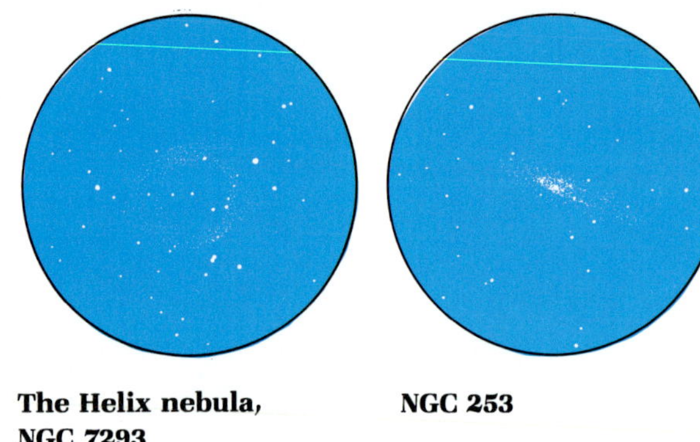

The Helix nebula, NGC 7293 — **NGC 253**

Eta's primary is 5.8, the secondary a magnitude dimmer, separated by 1.7 arcseconds. Although it has only one-third the separation of gamma's components, the stars are more evenly matched. Which is easier for your telescope, gamma or eta?

Sculptor, the Sculptor's Tools

Sculptor, an otherwise obscure southern constellation, lies near the south pole of the Galaxy, free of obscuring galactic dust, and boasts one of the brightest and most impressive southern spirals, NGC 253. With a small telescope, you will see it as an elongated glow; with a large telescope, you will be able to see the mottled surface, which photographs show as tightly wound spiral arms with complex dust lanes.

Only 1° southeast of this seventh-magnitude galaxy lies an eighth-magnitude globular cluster, NGC 288. While you're galaxy-hunting, don't miss NGC 55, a large nearly edge-on spiral and the brightest of the Sculptor galaxies, and NGC 300, another member of the same group.

Grus, the Crane

This constellation, representing a crane or wading bird, is too low in the south for serious observation.

Phoenix, the Phoenix

Representing the mythical bird that rose renewed from the ashes of its funeral pyre, Phoenix is a recent constellation. The Arabs knew the stars as the Boat, moored beside the River Eridanus.

The stars in star chart 20 are highest at:

July 23 at 4 A.M.; August 7 at 3 A.M.; August 23 at 2 A.M.; September 7 at 1 A.M.; September 22 at midnight; October 7 at 11 P.M.; October 23 at 10 P.M.; November 7 at 8 P.M.; November 23 at 7 P.M.

N

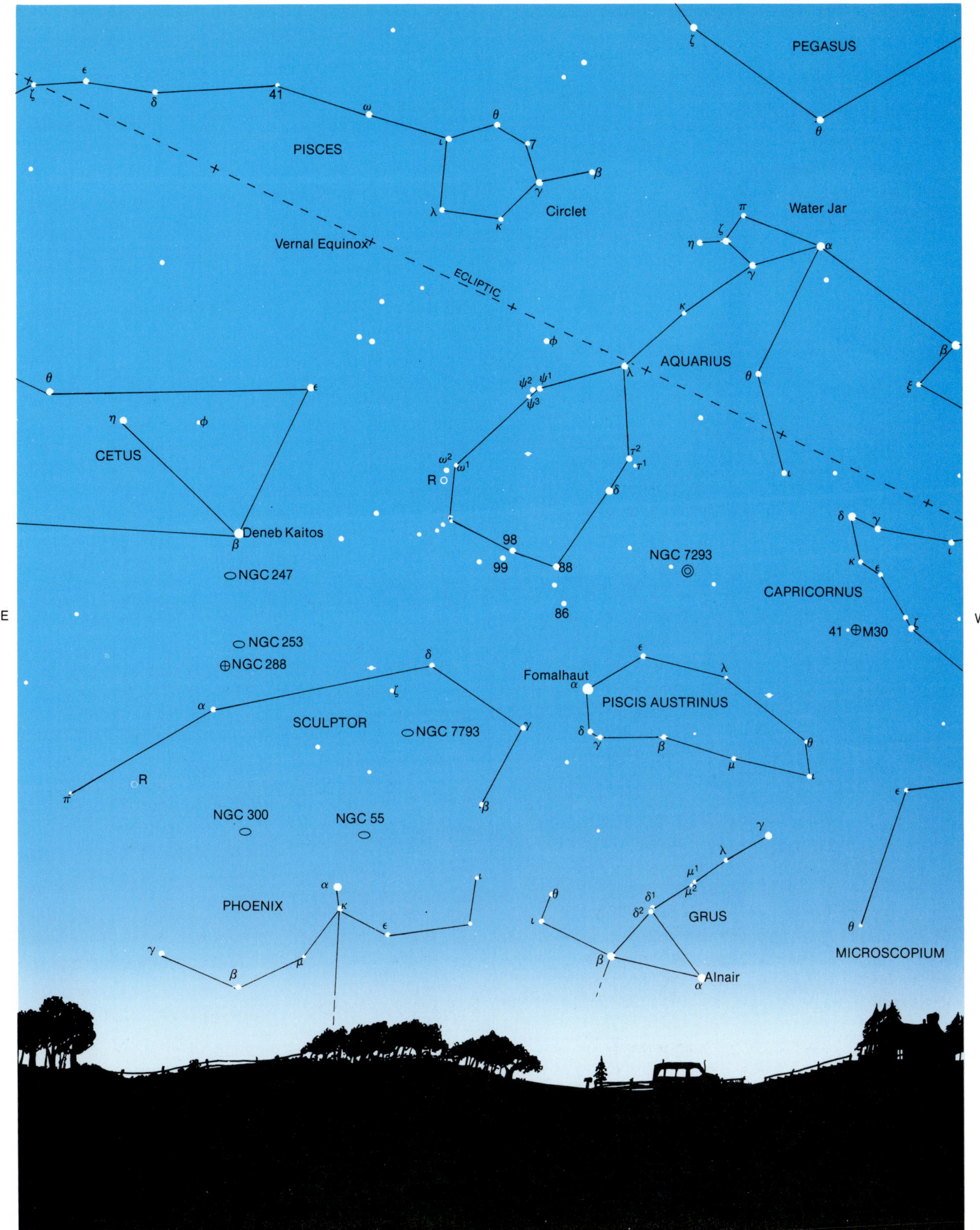
PEGASUS
PISCES
Circlet
Vernal Equinox
ECLIPTIC
Water Jar
AQUARIUS
CETUS
Deneb Kaitos
NGC 247
NGC 7293
CAPRICORNUS
41 M30
NGC 253
NGC 288
Fomalhaut
PISCIS AUSTRINUS
SCULPTOR
NGC 7793
NGC 300
NGC 55
PHOENIX
GRUS
Alnair
MICROSCOPIUM

E

W

S

STAR CHART 21
Facing north

Cassiopeia dominates the sky overhead in the late fall. This bright, quite unmistakable starry zigzag resembles the letter *M* above the north celestial pole, as you face north. (Of course, if you face south, Cassiopeia resembles the letter *W*. It all depends on your point of view.) Trace the soft glow of the Milky Way from Cassiopeia toward the setting stars of late summer, through Cepheus, Cygnus, and Aquila (star chart 17); or follow it into the rising winter sky, past Perseus, Auriga, Gemini, and Taurus (star chart 1).

Cassiopeia, the Queen

Like daisies in a field, you'll find open clusters scattered along the Milky Way in Cassiopeia. But the two best open clusters associated with Cassiopeia aren't even in Cassiopeia—they're h and chi (χ) Persei just across the official constellation boundary, in Perseus. The best way to find them is to start at gamma (γ) Cassiopeiae, hop to delta (δ), then hop again one and a half times that distance to the twin clusters. (See star chart 22 for more information.)

You can find M52 almost exactly the same way. Start from alpha (α) Cassiopeiae, hop to beta (β), then go one and a quarter times their distance. M52 is a pretty cluster of a hundred stars. A rather similar cluster lies right next to the star phi (φ). It's a slightly brighter cluster than M52, and particularly easy to find with binoculars. If you like searching out clusters, you don't have to stop yet: NGC 654, NGC 663, and M103 lie side-by-side between delta (δ) and epsilon (ε). M103 has two dozen stars; NGC 654 and NGC 663 have three times as many. These clusters can be hard to find, scattered as they are in the rich star fields of the Milky Way.

That's not all the clusters in Cassiopeia. With large binoculars or a telescope using very low magnification, look for NGC 129, NGC 225, IC 1905, and NGC 7243, bright clusters with a few dozen stars each. NGC 7789 is quite different. As bright as the other clusters, it contains several hundred stars, all quite faint. Because of the background stars, these objects are virtually invisible unless you take a broad view.

Among the clusters in Cassiopeia there are nebulae, but most of them are large and faint. If you

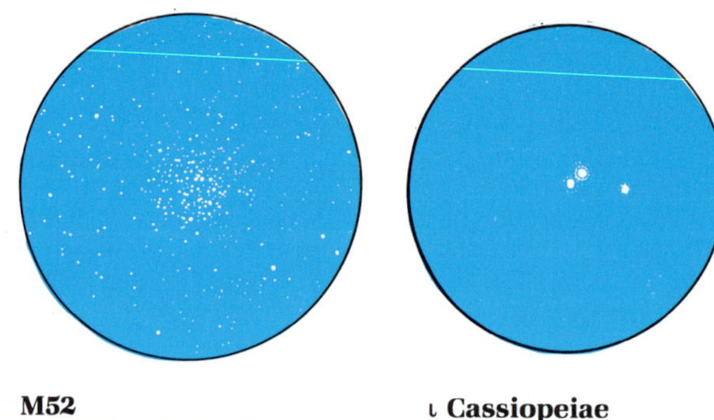

M52 **ι Cassiopeiae**

are under dark skies with a large telescope, see if you can find the Bubble nebula, NGC 7635, when you next observe M52.

Eta (η) Cassiopeiae is a beautiful double. The third-magnitude primary is yellow-white like our sun, but the seventh-magnitude secondary is deep orange in color. They lie 12 arcseconds apart. A large telescope will show you the color difference better than a small one can.

Iota (ι) is a triple. The components are 4.6, 6.9, and 8.4 magnitude. The primary and secondary are 2.5 arcseconds apart, orbiting each other with a period of 840 years, while the third star lies 7 arcseconds distant.

Our challenge star is lambda (λ), an exceedingly close double. With a good 8-inch telescope on a night of excellent seeing, you should be able to tell it's a double. The star's blue-white components are nearly matched—5.3 and 5.6 magnitude—but they lie a mere 0.6 arcseconds apart! Even if you can't split them, the star image should appear slightly elongated.

In 1572, a star as bright as Venus appeared one night in Cassiopeia, and was seen by the great Danish astronomer Tycho Brahe. The star was a supernova; it blew itself to bits. The remnant of the star is a radio source four hundred years later. We're due for another supernova—overdue in fact—so keep looking.

The stars in star chart 21 are highest at:

August 23 at 4 A.M.; September 7 at 3 A.M.; September 22 at 2 A.M.; October 7 at 1 A.M.; October 23 at midnight; November 7 at 10 P.M.; November 23 at 9 P.M.; December 7 at 8 P.M.; December 23 at 7 P.M.; January 7 at 6 P.M.

S
N
W
E
PEGASUS
Sirrah
M33
TRIANGULUM
Mirach
R
M32
M31
NGC 205
NGC 7662
ANDROMEDA
NGC 891
M34
Algol
PERSEUS
51
6
11
LACERTA
R
NGC 7789
2
5
4
CASSIOPEIA
NGC 7243
NGC 457
h
Algenib
Caph
NGC 129
M103
NGC 663
NGC 225
NGC 654
IC 1905
NGC 7635
M52
Nova 1572
Stock 23
CAMELOPARDALIS
7
NGC 1502
CEPHEUS
24
31
11
T
Polaris
NCP
DRACO
NGC 2403
NGC 6543
27
URSA MINOR
DRACO

STAR CHART 22 Overhead, facing south

The bright stars of winter are well up as the last of the fall stars pass overhead. Aries, the Ram, first constellation of the zodiac, follows Pisces, last of the zodiac figures, starting the eternal cycle yet again. Perseus and Andromeda play hero and heroine, while the sea monster Cetus rages ominously below.

Perseus, the Hero

The northern Milky Way is rich in open clusters —and Perseus certainly has its share. Most famous is the Double Cluster, h and chi (χ) Persei. Since these clusters are visible to the naked eye on a clear, dark night midway between Perseus and Cassiopeia, these spectacular clusters were given names appropriate for stars.

Perseus has another open cluster so obvious that it's easy to miss. Do you see all the bright stars around alpha (α)? They're called the Perseus Moving Cluster because they're moving together through space, just like a star cluster. There are roughly fifty stars all told in this group. Barely visible to the naked eye, but quite bright through a binocular, is the cluster M34. Compare it with h and chi.

Near xi (ξ) is a large, exceedingly dim emission nebula named after the state of California, which its outline resembles. Xi is a young, hot, blue star; its light makes this cloud of interstellar gas glow.

Keep an eye on beta (β), the famous winking star, Algol. Algol is a double that's much too close to split with a telescope. Every 2 days, 20 hours, 49 minutes, the faint star eclipses the bright star, and we see Algol go from 2.1 magnitude to 3.4 magnitude and back in about ten hours. When beta is faintest it matches rho (ρ) in brightness. Watch it nightly until you catch it in action!

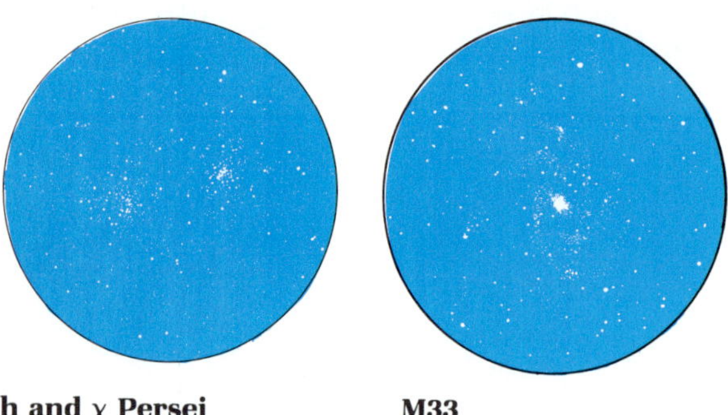

h and χ Persei **M33**

Aries, the Ram

Aries lies on the zodiac, and once marked the spot where the north-moving sun crossed the celestial equator at the beginning of spring. No more. The slow wobbling (precession) of earth's axis has moved "the first point of Aries" into Pisces, near the Circlet (see star chart 19).

Gamma (γ) is a stunning double star consisting of two equally bright stars lying 8 arcseconds apart. With a small telescope at high magnification it resembles the headlights of a distant automobile. Gamma Virginis is similar, but its stars are closer, and yellowish in color.

Triangulum, the Triangle

Just a little off the line between alpha (α) Trianguli and beta (β) Andromedae is M33, the Triangulum galaxy. A member of the Local Group of galaxies, M33 is a loose spiral 1° long by 40 arcminutes wide. Oddly enough, it is usually easier to see M33 with a binocular than with a telescope because it's so large and faint. If you can spot M33 naked-eye, you've got a very good night and a very good observing site.

Iota (ι) is a pretty gold and blue double. The primary is 5.3 magnitude and the secondary is 6.9 magnitude. They lie 3.9 arcseconds apart.

Pisces, the Fishes (see also star chart 19)

The eastern end of Pisces offers a galaxy and several doubles. The galaxy is M74, a face-on spiral galaxy. It's a small version of M33, one-sixth as large. With small telescopes, it is little more than a dim glow in the eyepiece.

Alpha (α) is a challenging double for small telescopes. The secondary shines at fifth magnitude, while the primary is fourth. The stars are currently getting closer—they lie 2 arcseconds apart—as they circle each other every 933 years.

The stars in star chart 22 pass overhead at:

September 7 at 4:30 A.M.; September 22 at 3:30 A.M.; October 7 at 2:30 A.M.; October 23 at 1:30 A.M.; November 7 at 11:30 P.M.; November 23 at 10:30 P.M.; December 7 at 9:30 P.M.; December 23 at 8:30 P.M.; January 7 at 7:30 P.M.; January 23 at 6:30 P.M.

N
E
W
S
CAMELOPARDALIS
NGC 1502
IC 1905
Stock 23
NGC 654
NGC 663
M103
NGC 225
NGC 129
Caph
NGC 7789
NGC 457
CASSIOPEIA
R
h
Capella
AURIGA
Algenib
PERSEUS
California Nebula
M34
NGC 891
Algol
ANDROMEDA
M31
M32
NGC 205
R
51
Mirach
Sirrah
TRIANGULUM
M33
The Pleiades
Aldebaran
The Hyades
TAURUS
Hamal
ARIES
ECLIPTIC
M74
PISCES
CETUS
Menkar
Alrisha
M77
Mira

STAR CHART 23
Facing south

While mighty Orion rises in the east and Pegasus sets in the west, the southern sky seems dull and starless at first. It is not. Though the constellations that occupy this dark region are lackluster, they are nonetheless interesting to telescopic observers. Naked-eye observers should, every time they even casually glance at this area, check on the remarkable variable star Mira, in the tail of the Whale.

Cetus, the Whale

We depict Cetus as a friendly whale swimming through the watery autumn heavens, but he has not always been seen as so benign. Imagine the circle of stars we show as the flukes as an evil head outstretched from a gross body wallowing in the waves—coming to devour the tender princess, Andromeda (see star chart 22). This large and rather dim figure lies well away from the Milky Way and also well away from the Virgo Cluster of galaxies, so it is poor in both galactic and extragalactic objects for telescopes. However, that doesn't mean it's not interesting.

With no optical aid whatsoever, you can follow the variable star Mira, omicron (ο) Ceti, the "miracle" star of Cetus. Mira is a long-period variable, a vast red giant star that pulsates with a period of 332 days. At its brightest, Mira shines as a second-magnitude star, but at its dimmest, it's tenth magnitude, totally invisible to the eye.

A star very much like our sun is in Cetus—tau (τ)—visible because it lies only 10.7 light-years away. If we were in orbit around tau Ceti, our sun would look just about like tau Ceti looks in our night sky.

For the galaxy hunter, there are M77 and NGC 247 to find. M77 is not particularly bright, but it's near delta (δ) and fairly easy to find with a medium-sized telescope. Look at it carefully. Do you see what's different about it? The nucleus is bright and star-like. Besides being unusually bright, the nucleus of this galaxy also emits radio waves. NGC 247 is a largish spiral galaxy best seen at low magnification. While you're in the area catch NGC 253, across the border of Cetus, in Sculptor (star chart 20).

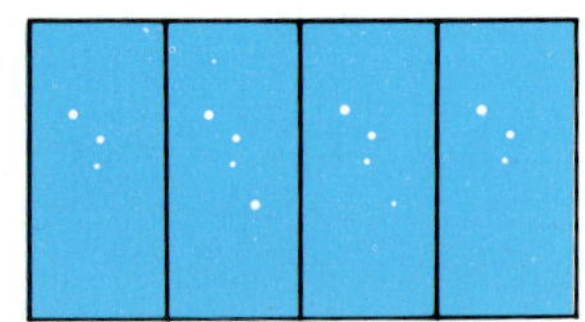

ο Ceti's variation

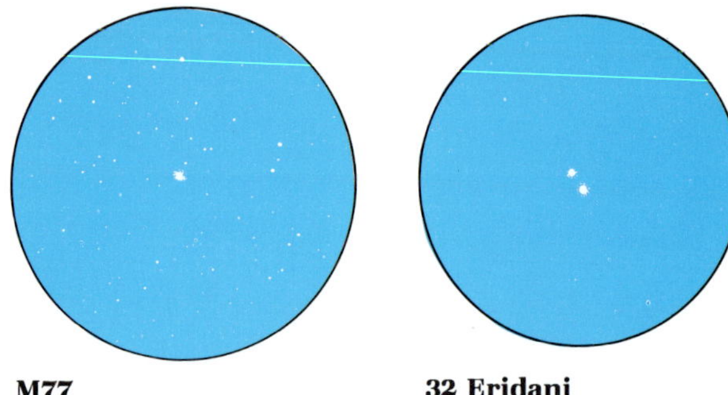

M77 **32 Eridani**

Fornax, the Furnace

Fornax is a miserable little constellation with hardly anything to recommend it. One faint member of our local group of galaxies lies in it, as does a distant cluster of galaxies, but these are objects for large observatory telescopes.

Eridanus, the River (see also star chart 3)

How far can you trace Eridanus? If you live south of 30° latitude, you can follow it all the way from its origin near Orion to its end at the first-magnitude star Achernar, below the horizon on these charts.

The star 32 Eridani is a nice double with a yellow primary and a blue-white secondary, magnitudes 4.8 and 6.1, respectively. They lie 6.8 arcseconds apart, fair game for almost any telescope. Theta (θ), also named Acamar, is another good double. The brighter of the two is third magnitude; the dimmer, fourth. They lie 8 arcseconds apart in the sky. Omicron-2 (o^2) is a wide double (83 arcseconds) with a ninth-magnitude white dwarf companion star. (Both Sirius and Procyon have white-dwarf companions, but omicron-2's is much easier to see.)

For observers with medium and larger telescopes, locate the planetary nebula NGC 1535. It's small—20 arcseconds across—so use enough magnification to enable you to distinguish it from a star. NGC 1300 is an elegant spiral galaxy often seen in textbook photos. Try for the real thing!

The stars in star chart 23 are highest at:

September 7 at 4:30 A.M.; September 22 at 3:30 A.M.; October 7 at 2:30 A.M ; October 23 at 1:30 A.M.; November 7 at 11:30 P.M.; November 22 at 10:30 P.M.; December 7 at 9:30 P.M.; December 23 at 8:30 P.M.; January 7 at 7:30 P.M.; January 23 at 6:30 P.M.

N

TAURUS

PISCES

ECLIPTIC

Menkar

Alrisha

M77

Mira

32

NGC 1535

CETUS

ERIDANUS

NGC 1300

πCet

NGC 247

E

W

NGC 253

NGC 288

FORNAX

SCULPTOR

R

NGC 300

NGC 55

PHOENIX

S

4. Exploring the Moon

The moon is by far the most prominent object in the nighttime sky, and it holds many rewards for both naked-eye and telescopic observation.

The moon is our planet's natural satellite. Its diameter is 2,160 miles, one-fourth of our planet's diameter, and it has 1/81 of earth's mass. It orbits our planet at an average distance of 238,000 miles. Sunlight illuminates it, and as it moves around us, we see it lit up in a regular series of phases. It takes 29.5 days to go through a complete cycle of phases.

Unlike the earth, which rotates on its axis 365 times during each revolution around the sun, the moon rotates on its axis only once during each of its revolutions around the earth. If the moon rotated faster, or not at all, it would turn what we call its "dark side" toward us. Because of its once-a-month rotation, the same side of the moon always faces the earth. Humans never knew what the far side of the moon looked like until spacecraft flying past it sent back photographs in 1959.

Like the sun, the moon appears to move through the sky from east to west because of the earth's daily rotation on its axis. But as it orbits the earth, the moon also travels 1/29th of the way around the sky each day, going from west to east. This causes it to rise a little less than one hour later each night.

The moon's orbit lies in nearly the same plane as that of the earth around the sun. This means that the moon's path through the earth's sky is nearly the same as the sun's path and the paths of the other planets. This path is called the *ecliptic*. The moon never strays more than 5° from the ecliptic path. The ecliptic is shown on the sky maps and star charts.

We are all accustomed to seeing the moon's different phases, ranging from a tiny crescent to a full moon. Many calendars show the dates of the moon's phases, which are called new moon, first quarter, full moon, and third quarter, or last quarter. The phases are easy to understand if you think of the moon circling the earth out in space, where the sun always shines. When the moon is between

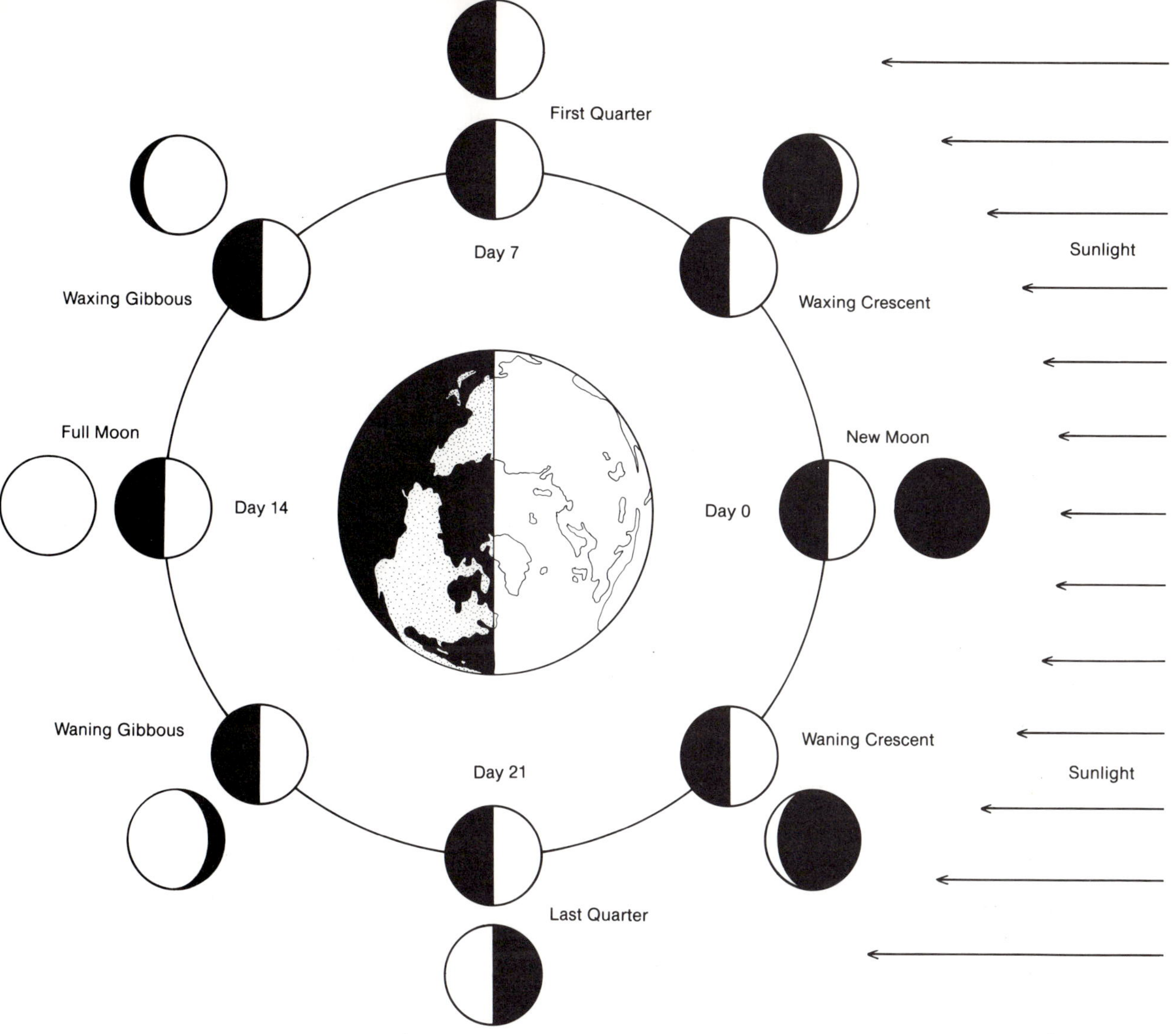

the earth and the sun, the lit side is away from us, and the moon rises at sunrise and sets at sunset, just as the sun does. This is the new moon, when we see no moon in the sky at night.

Fourteen days later the moon has circled to the opposite side of the earth from the sun. This means that the entire face of the moon we see is lit by sunlight, and that the moon rises at dusk, passes its highest place in the sky at midnight, and sets at dawn. This is the full moon.

As the moon passes from new moon to full moon, its visible side grows from a tiny crescent to the full face, and it rises later in the day. This means that when the sun sets we find the moon already up in the sky, and it sets sometime during the night. At first quarter, for instance, seven days after new moon, half of the moon is vis-

The phases of the moon. As the moon orbits earth, we watch its face pass through phases, beginning with new moon, when we see only the shadowed side; through first quarter, when we see it half lit; to the fully lit full moon; then to last quarter; and finally back to new moon.

ible, it appears high overhead at dusk, and it sets about halfway through the night.

After full moon, the moon continues around the earth. The illuminated part of the moon decreases from a full face to a sliver, and it rises later in the night. This means that there is no moon at dusk. It rises later each night, and is still in the sky when the sun rises. At third quarter, twenty-one days after new moon, we again see half the moon, but it doesn't rise until about halfway through the night, and is high in the sky at dawn.

Using your telescope, look at the moon a few days after new moon. You'll see it as a thin crescent setting an hour or two after sunset. Enclosed in the crescent you'll see the rest of the lunar disk glowing faintly, lit by light reflected from the earth. In the lit area, you'll notice several craters and Mare Crisium ("The Sea of Crises") thrown into high shadow relief. The craters are impact scars, places where asteroids miles in diameter slammed into the lunar surface. Most of the ones you'll see are billions of years old. Lunar maria (Latin for "seas") are large flat areas so named by seventeenth-century moon mappers who thought they were filled with water. They are actually vast plains of lava that poured from the moon's interior early in its history.

A few nights later, as the moon continues on its path around the earth, it will be higher in the sky at sunset and stay up for a few hours. Now you'll have a chance to examine it at your leisure. The crescent will be much wider each night as the moon "waxes," or becomes more fully illuminated. The craters along the edge of the shadow line (which is called the "terminator") will really knock your eyes out. Some of them will be right on the terminator, filled with inky-black shadow. Others will be farther back and you'll see into them.

Mare Crisium, the "sea" you saw when the moon was a crescent, is now lit by sunlight from a higher angle so that it no longer shows towering, shadowed walls. Other maria are visible, too. Mare Tranquillitatis (where men first stepped on the moon), Mare Nectaris, and Mare Foecunditatis lie under lower lighting. If you look closely at them you'll see that they even have a few small craters on them. Near the terminator you'll see long, low ridges in their surfaces. These "wrinkle ridges" formed as the lava that made the seas poured out, then cooled, shrank, and wrinkled.

The "seas" and craters of the moon

As you look at the cratered highlands, note that some craters are surrounded by whitish streaks that go long distances across the moon. These lunar rays are made of material thrown out across the surface, where it landed in long streaks, when the moon was struck. Rays mark the youngest craters on the lunar surface.

Seven nights after new moon, the moon reaches first quarter. At sunset, you'll see it high in the south. The terminator now runs right down the middle of the visible side of the moon. The names of the lunar features you can see in this phase are found on the map, and with half the moon showing, it's easy to find your way around. Don't miss the great mountain arc of the lunar Apennines surrounding the Mare Imbrium. Be sure to locate the beautiful crater chain of Ptolemaeus, Alphonsus, and Arzachel.

The night after first quarter, find Tycho, a new bright crater in the southern highlands, and Clavius, a great ringed plain with many smaller craters tucked inside. At high magnification, observe the Straight Wall, a crack in the lunar crust, as well as Bessel and Linné, craters on Mare Serenitatis.

As the moon waxes to gibbous (nearly full), the ray systems become brighter and more obvious. Watch, from night to night, how craters that were starkly illuminated the night before flatten and disappear in the brilliant glare of the sun. Observe the huge expanse of Mare Imbrium, ringed by mountains, and the flooded craters along the shores of Oceanus Procellarum and Mare Nubium. And don't miss the big crater Copernicus. Dramatic lighting on the first night it becomes visible makes Copernicus a sight you'll want to show your friends.

When the moon is full it is on the far side of the earth from the sun. This means that the side we see catches the sun's light full on its face. There are no visible shadows on the surface of the full moon, making it something of a disappointment to those who expect it to be the best phase for observing. Instead of deep-shadowed craters, at full moon you'll see bright rings marking their outer walls. Many small craters look like tiny white spots. But the rays are easiest to see at full moon. The systems of rays from Tycho and Copernicus are among the most striking lunar features at this phase.

After full moon, the moon rises later and later each night, and goes through its phases in reverse order: gibbous, last quarter, waning crescent, and finally new. You might want to set your alarm clock to wake up in the wee hours before dawn so you can watch the evening shadows on craters you saw during the lunar morning.

About once a year, on the average, you will be able to see an eclipse of the moon. Eclipses occur when the moon's path takes it through the shadow that our planet Earth casts in space. It takes a few hours for the moon to pass through the earth's shadow. Watch the monthly astronomy magazines for information on upcoming lunar eclipses.

5. Exploring the Planets

The Bright Planets *Four of the planets are bright and easy to find with your unaided eyes and with your telescope: Jupiter, the giant planet; Saturn, the ringed planet; Mars, the red planet; and Venus, the cloudy planet. Because the planets move—indeed, the word* planet *means "wanderer"—they can't be plotted on the maps or charts. However, you will always find them near the ecliptic, which is included on the charts and maps.*

JUPITER

Jupiter looks like a bright, yellowish, slightly flattened ball. Even with a telescope, you probably won't see anything on Jupiter at first, but you will notice several small bright "stars" strung out in a line near it. These are the Galilean satellites, the very same that Galileo discovered with his tiny telescope in 1610. Each night that you observe, you'll see the satellites in a different order because the four of them are continually circling Jupiter with periods ranging from 1.7 to 17 days. Sometimes you won't see all four because one or more will be hiding in Jupiter's shadow, or behind or in front of Jupiter itself.

With a telescope, look for the planet's cloud belts. They appear in pastel shades of gray, khaki, rust, and brown. Look for the famous Great Red Spot, which varies from pale pink to rust in color. If you don't see it one night, look another night. It may have been on the other side of the planet. Jupiter's disk is 45 arcseconds across, varying somewhat depending on where Earth and Jupiter happen to be in their orbits.

SATURN

Saturn looks oval or spindle-shaped with binoculars. Through small telescopes, it appears as a tiny round planet encircled by a ring. The most impressive thing about Saturn, perhaps, is how crisp and delicate it looks shining in the eyepiece. Saturn is eight times larger than Earth and its distance averages just under one billion miles from us. A magnification of 60× is sufficient to reveal the planet's rings.

Near Saturn, you can usually see its biggest satellite, Titan, which

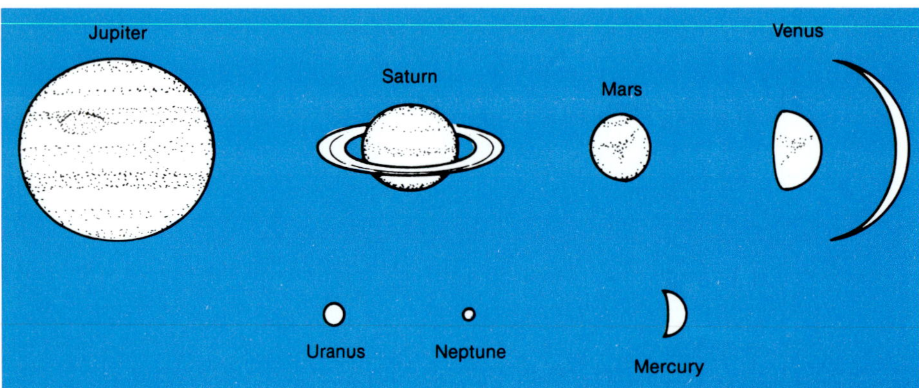

The planets through a telescope. With this book held at arm's length, the planets on this page appear as large as they will at 160× magnification through a medium-sized telescope. Mars is shown the size it appears when it is closest to Earth for a few weeks every two years; it usually appears smaller. Venus is shown in two different phases. Jupiter and Saturn are fairly constant in size and usually offer the best viewing for small telescopes.

looks like a small star. (Under good conditions, you might glimpse several other satellites looking like faint stars.) If you look for Titan regularly, you can follow it as it revolves around Saturn every 16 days. A series of little sketches made each night when you observe will help you spot its motion.

MARS

To telescopic observers, the red planet Mars is interesting for only three months out of every two years, when it's closest to Earth. At those times, called "opposition" because the planet is then opposite the sun in the sky, you can see features on the planet's surface. Look for grayish or greenish markings on the disk of the planet. Prominent markings like Syrtis Major have been observed now for hundreds of years. Around the turn of the century, astronomer Percival Lowell reported seeing networks of very fine lines he thought were canals built by Martians, but spacecraft have shown the planet's surface is ancient, dusty, and lifeless.

When Mars is near opposition, its disk may be as large as 25 arc-seconds. Outside that three-month period, Mars is so far away that you will see very little with a telescope but the planet's distinctive ruddy color and a planetary disk.

VENUS

Venus is the brightest of the planets. It outshines everything in the sky but the sun and moon. Because the orbit of Venus is smaller than Earth's orbit, we always see it in the sky near the sun; hence Venus is known commonly as the "evening star" or "morning star."

Venus goes through phases similar in shape to the phases of the moon. However, Venus also changes in apparent size. When this planet appears as a thin crescent, it is near us in its orbit and as large as 60 arcseconds across. When it is in its "half-Venus" phase, the planet is at some intermediate distance, appearing 20 arcseconds across. When the disk is full or gibbous, the planet is on the far side of its orbit, and appears only 5 arcseconds in diameter.

Chances are you'll see little or no detail of any kind on Venus, but that doesn't mean it's not worth looking at. Follow Venus through its phases, especially as it swings toward or away from superior conjunction (when it passes between Earth and the sun).

The Elusive Planets

Mercury, Uranus, and Neptune are much harder to find than the bright planets. Mercury circles close to the sun, so it can be seen for only a few weeks during morning twilight or a few weeks during evening twilight four or five times each year. Uranus and Neptune are both too faint to be seen without a telescope, although almost any telescope is powerful enough to spot them both—if you know just where to look.

MERCURY

Mercury is sufficiently difficult to find that some famous professional astronomers have never seen it. But you should have no trouble if you'll make the effort to look in the right place at the right time. Mercury will appear in the morning or evening twilight sky as a bright, slightly yellowish or reddish star, close to the horizon. The bright wide field and large light-gathering power of a binocular will allow you to search the horizon sky and find Mercury even if thin clouds, city haze, or approaching dawn make naked-eye detection difficult.

URANUS

Uranus is a giant planet, like Jupiter and Saturn. Unlike them, however, Uranus is too faint (sixth magnitude) to see with the naked eye. With a suitable telescope and chart for the year you are observing, though, you'll have no problem locating this planet among the stars.

Once you think you see Uranus, switch to the highest power eyepiece you have. Carefully inspect the suspect and compare it to stars of the same brightness. The planetary disk will appear quite small—4 arcseconds in diameter—but under good conditions its image will look distinctly fatter than a star image. Uranus's greenish tinge, which is fairly subtle, is an additional tip-off that you've really found another planet!

NEPTUNE

Neptune is still farther from the sun than Uranus, and is somewhat fainter (eighth magnitude) and somewhat smaller (2.5 arcseconds). It, too, requires a finder chart, some patience, and high power for absolute confirmation that you've found it. While there isn't much to see on these outer denizens of the solar family, simply finding them there and seeing them with your own eyes (with a little help from your telescope, of course) will give you great satisfaction.

PLUTO

Pluto is so faint (fourteenth magnitude) and so distant that few amateur astronomers have seen it. Pluto requires good charts, lots of patience, and a 10-inch or larger telescope to find.

Solar System and Planet Facts

	Distance[1]	Orbital Period	Diameter[2]
Sun	———	———	———
Mercury	0.3871	88 days	3,014 miles
Venus	0.7233	225 days	7,545 miles
Earth	1.0000	365 days	7,927 miles
Mars	1.5237	687 days	4,220 miles
Jupiter	5.2028	11.86 yrs	89,000 miles
Saturn	9.5388	29.45 yrs	74,580 miles
Uranus	19.1914	84.07 yrs	32,500 miles
Neptune	30.0611	161.81 yrs	30,200 miles
Pluto	39.5294	248.00 yrs	1,500 miles

1. In astronomical units of 92,956,000 miles, or 149,598,000 km.
2. To change to kilometers, multiply by 1.609.

6. Learning More About Astronomy

With the aid of this book, you'll taste the delights of stargazing. Then what? What's going on in the sky this month? Are there any good books on astronomy? Where can you see a planetarium show? Are there any observatories open to the public? Here are some hints on going further.

MONTHLY ASTRONOMY MAGAZINES

If you're new to stargazing, *Astronomy* magazine is the best for you. Its colorful, lavishly illustrated feature articles are written in nontechnical language you can understand, and its monthly departments on equipment, observing, photography, and what you can see in the sky are very useful. *Astronomy*'s calendar of upcoming events will help you meet like-minded people. Of course, you'll soon be scanning the advertisements for telescopes—in every size, type, and price range. *Astronomy* is available on many newsstands, in most public libraries, or by subscription ($18 per year) from the Kalmbach Publishing Co., 1027 N. Seventh Street, Milwaukee, WI 53233.

Another worthwhile monthly is *Sky & Telescope*. Since it's written for technically inclined and advanced amateur astronomers, beginners often find this magazine tough going. If you're working in a technical field, though, you'll probably like it. This magazine features current research news, semitechnical papers for advanced amateur astronomers, and columns on telescope making and deep-sky observing. You'll find it on specialty newsstands and in some big-city public libraries.

CONSTELLATIONS AND STAR LORE

Chet Raymo's *365 Starry Nights* (Englewood Cliffs, NJ: Prentice-Hall, 1982) is a thoroughly charming introduction to stargazing and astronomy. Every night of the year has its own lesson, and it's illustrated with the author's delightful drawings.

For a direct sampling of ancient starlore, you can't beat Stanley Lombardo's *Sky Signs: Aratus' Phaenomena* (Berkeley: North Atlantic Books, 1983). This edition translates the Greek poet Aratus's 2,200-year-old song of the constellations into modern English.

Finally, there's *Star Names—Their Lore and Meaning*, by Richard H. Allen (New York: Dover, 1963). Knowing where the names come from helps in learning them, and this book will give you a feel for the diversity of cultural influences that have shaped astronomy.

SKYGUIDES AND ATLASES

Sky guides come in every size and level of complexity. This book, for example, is a basic, nontechnical introduction to stargazing. A bit more sophisticated, but still great for beginners, is a tiny book called *The Night Sky*, by Ian Ridpath and Wil Tirion (London: Collins, 1985). Only 3¼ inches by 4½ inches and 240 pages thick, it's a fine companion at the telescope.

Larger, heavier, and more detailed, the *Universe Guide to Stars and Planets* by Ridpath and Tirion (New York: Universe Books, 1985) shows more stars and more deep-sky objects.

The top of the field is *A Field Guide to the Stars and Planets* (Peterson Field Guide Series) by Donald H. Menzel and Jay M. Pasachoff (Boston: Houghton Mifflin, 1983). This 483-page "field guide" packs so much into its pages that it's hard to find anything, but it's complete and authoritative.

Among star atlases, Wil Tirion's *Sky Atlas 2000.0* (New York: Cambridge University Press, 1981) is the best. Twenty-six large charts cover the whole sky, showing 43,000 stars down to eighth magnitude, plus 2,500 deep-sky objects. It's a clear, uncluttered star atlas simple enough to use with your first telescope, but good enough to satisfy you for many years.

If you crave "data," then *Sky Catalogue 2000.0*, Vol. 2, *Double Stars, Variable Stars, and Nonstellar Objects*, by Alan Hirshfeld and Roger W. Sinnott (New York: Cambridge University Press, 1983) is for you. It's a big book packed with numbers—the latest data on thousands of deep-sky objects.

ASTRONOMY AS A HOBBY

Astronomy can seem rather daunting as a hobby—until you read Ken Fulton's magnificent book *The Light-Hearted Astronomer* (Milwaukee: AstroMedia, 1984). With a delightful mixture of humor and good common sense, Fulton tells you how to survive "astronomy's jungle," how to cope with "telescope fever," and much, much more. It's offbeat and highly recommended.

For a handy mix of practical observing hints and the science of astronomy, find a copy of *The Universe Next Door* by Terry Holt (New York: Scribner's, 1985). If you want to know more about what celestial objects *are* but don't want to read a dull textbook, you'll like Holt's book.

For the do-it-yourselfer, my own book *Build Your Own Telescope* (New York: Scribner's, 1985) provides complete plans for five telescopes that anybody can build, from a simple 4-inch reflector to a 6-inch refractor and 10-inch reflector, plus practical down-to-earth advice on every aspect of telescopes.

PLANETARIUMS AND MUSEUMS

Planetariums are educational institutions and they welcome visitors. They have daily "sky shows" featuring stars and constellations realistically projected onto a hemispherical dome, and exhibits about astronomy. Their bookstores are often the best places to find astronomical magazines and books about astronomy. This list is just a small sample: There's a planetarium in practically every major city in the United States. Call in advance for show times.

Atlanta: *Fernbank Science Center Planetarium*, 156 Heaton Park Drive, Atlanta, GA 30307 (404-957-9653). Just east of Atlanta, Fernbank has a 36-inch reflecting telescope as well as a planetarium show.

Los Angeles: *Griffith Observatory*, 2800 East Observatory Road, Los Angeles, CA 90027. Located in Griffith Park. Open 2–10 P.M. Tuesday to Friday; Saturdays 10:30 A.M.–10 P.M.; Sunday 1–10 P.M.; also open Mondays during the summer. A complete facility, with large refractor telescopes, daytime solar viewing, exhibits, bookshop, planetarium.

Rochester: *Strasenburgh Planetarium*, 663 East Avenue, Rochester, NY 14603 (716-244-6060). A planetarium with a reputation for first-rate shows on a large dome. Exhibits and shop.

San Francisco: *Chabot Observatory and Planetarium*, 4917 Mountain Blvd., Oakland, CA 94619. A planetarium and telescopes—

including a 20-inch refractor—available for public viewing every Friday and Saturday beginning at 7:30 P.M.; also offers workshops in building telescopes.

Vancouver: *Macmillan Planetarium,* Vanier Park, 1100 Chestnut Street, Vancouver, BC, V6J 3J9 Canada (604-736-4431). Enjoys a reputation for giving one of the best "big dome" sky shows in North America.

Washington: *Albert Einstein Spacarium,* National Air and Space Museum, Washington, DC 20563 (202-381-4193). The emphasis is on spaceflight, and the kids will love it. See America's pioneering spacecraft, and touch a moon rock.

A UNIQUE SIGHT

If you're on the road out west, why not stop in to see the **Barringer Meteor Crater**? It's located 19 miles west of Winslow, Arizona, on I-40, and is open from 8 A.M. to sundown. Hike around the rim of a mile-diameter meteor crater blasted into the ground about 10,000 years ago. The admission fee is rather high—but where else can you tour a meteor crater?

ASTRONOMICAL OBSERVATORIES

Observatories are research laboratories for astronomers, so public access to telescopes is severely limited or discouraged at night. However, many observatories do offer daytime tours during the summer months. Write for information in advance, as visitors' hours may change.

Allegheny Observatory, 159 Riverview Avenue, Pittsburgh, PA 15214. Located on a bluff above Pittsburgh, this beautiful older observatory has undergone recent modernization, and the 30-inch Thaw refractor now searches for planets circling distant stars. Open Monday–Saturday nights between April 1 and November 1, for lectures, for movies, and for viewing with the observatory's 13-inch refractor.

Kitt Peak National Observatory. Located two hours' drive south of Tucson, Arizona, off State Highway 86, at an elevation of 7,000 feet. Open 10 A.M.–4 P.M. every day except Christmas. There are more telescopes on this mountain than at any other site in the world. Largest is the 158-inch Mayall reflector, devoted to studying dim objects at the outermost reaches of the universe; equally impressive is the 300-foot-long McMath solar telescope devoted to our own sun.

McDonald Observatory, Mount Locke, Fort Davis, TX 79734. Drive north from Fort Davis or Kent on SR-118. Has 82-inch and 107-inch reflecting telescopes. Open Monday to Saturday 9 A.M.–5 P.M.; Sunday and holidays 1–5 P.M.; self-guided tour of the 107-inch telescope dome; lectures on weekdays and weekends; bookshop and exhibit center.

Palomar Mountain Observatory. Located about two hours' drive south of Los Angeles. Follow U.S. 395 up into the mountains, then turn on to SR-76. Open 9 A.M.–5 P.M. all year. This is the site of the world's largest effective reflecting telescope, the 200-inch Palomar telescope. There is a small exhibit building, and you can see the 200-inch telescope from a glass room on one side of the observing floor. There are two Forest Service campgrounds a few miles south of the observatory and many others nearby.

Yerkes Observatory, Williams Bay, WI 53191. Near popular resort town of Lake Geneva, two hours from Chicago. Houses the world's largest refracting telescope, the 40-inch donated by Charles Yerkes. This observatory has magnificent grounds and buildings. Open June 1 to September 30 on Saturdays 1:30–3:00 P.M.; lecture and tour every half-hour.

List of Constellations

Name	Pronunciation	Meaning	Abbr.	Notes	Month Best Seen	Star Chart#
Andromeda	an DROM a duh	the chained lady	And	A	November	19
Antlia	ANT lee uh	air pump	Ant	S	—	6
Apus	AY pus	bird of paradise	Aps	S	—	-
Aquarius	uh QWAR ee us	water bearer	Aqr	AZ	October	18,20
Aquila	AK wil uh	eagle	Aql	AG	August	17
Ara	AY ruh	altar	Ara	S	—	-
Aries	AIR eez	ram	Ari	AZ	December	22
Auriga	aw RYE guh	charioteer	Aur	AB	January	1
Boötes	bow OH teez	herdsman	Boo	AB	June	11
Caelum	KI lum	chisel	Cae	S	—	-
Camelopardalis	cuh MEL oh PAR duh lus	giraffe	Cam	C	January	4
Cancer	KAN sir	crab	Cnc	AZ	March	5
Canes Venatici	KAY neez VEN at ih see	hunting dogs	CVn		May	9
Canis Major	KAY niss MAY jor	greater dog	CMa	ABG	February	3
Canis Minor	KAY niss MY nor	lesser dog	CMi	A	February	5
Capricornus	CAP rih COR nus	sea goat	Cap	AZ	September	18
Carina	cuh RYE nuh	keel	Car	S	—	-
Cassiopeia	CAS ee oh PEE uh	the queen	Cas	ABCG	November	21
Centaurus	sen TOR us	centaur	Cen	BS	May	12
Cepheus	SEE fee us	the king	Cep	ACG	November	16
Cetus	SEE tus	whale, monster	Cet	A	December	23
Chamaeleon	shuh MAY lee on	chameleon	Cha	S	—	-
Circinus	SIR sin us	compass	Cir	S	—	-
Columba	co LUM buh	dove	Col	S	February	3
Coma Berenices	CO muh BER uh NI ceez	hair of Berenice	Com		May	7,10
Corona Australis	cuh ROW nuh aw STRAY lus	s. crown	CrA	S	August	14,15
Corona Borealis	cuh ROW nuh BOR ee AL us	n. crown	CrB	A	June	11
Corvus	COR vus	crow	Crv	A	March	8
Crater	KRAY tair	goblet	Crt	A	April	8
Crux	KRUKS	cross	Cru	BS	—	-
Cygnus	SIG nus	swan	Cyg	ABG	September	17
Delphinus	del FEE nus	dolphin	Del	A	September	17
Dorado	dor AY doe	gold fish	Dor	S	—	-
Draco	DRAY coe	dragon	Dra	AC	July	16
Equuleus	ee QWU lee us	foal	Equ		September	17
Eridanus	ih RID un us	river	Eri	A	December	3,23
Fornax	FOR nax	furnace	For	S	December	23
Gemini	JEM in ee	the twins	Gem	ABGZ	February	1
Grus	groose	crane	Gru		October	18,20
Hercules	HER cue LEES	the hero Hercules	Her	A	July	13
Horologium	HOR oh LOW gee um	clock	Hor	S	—	-
Hydra	HY druh	water snake	Hya	A	March	6,8,12
Hydrus	HY drus	small snake	Hyi	S	—	-
Indus	IN dus	American Indian	Ind	S	—	-

Name	Pronunciation	Meaning	Abbr.	Notes	Month Best Seen	Star Chart#
Lacerta	luh SIR tuh	lizard	Lac		March	19
Leo	LEE oh	lion	Leo	ABZ	March	7
Leo Minor	LEE oh MY nor	lesser lion	LMi		March	7
Lepus	LEE pus	hare	Lep	A	January	3
Libra	LEE bruh	scales	Lib	AZ	June	12
Lupus	LU pus	wolf	Lup	S	June	12,14
Lynx	links	lynx	Lyn		January	4
Lyra	LIE ruh	lyre	Lyr	AB	July	13
Mensa	MEN suh	table	Men	S	—	-
Microscopium	MY kro SCO pee um	microscope	Mic	S	September	18
Monoceros	muh NOS er us	unicorn	Mon	G	February	5
Musca	MUS cuh	fly	Mus	S	—	-
Norma	NOR muh	level	Nor	S	—	-
Octans	OCK tans	octant	Oct	S	—	-
Ophiuchus	OH fee U kus	serpent handler	Oph	A	July	14
Orion	oh RYE on	the hunter	Ori	ABG	January	2
Pavo	PAY voh	peacock	Pav	S	—	-
Pegasus	PEG uh sus	the flying horse	Peg	AB	October	19
Perseus	PER see us	Perseus	Per	AC	November	22
Phoenix	FEE nix	Phoenix	Phe	S	—	20,23
Pictor	PIK tor	easel	Pic	S	—	-
Pisces	PIE ceez	fishes	Psc	AZ	November	19,22
Piscis Austrinus	PIE sis aw STREE nus	s. fish	PsA	AS	October	20
Puppis	PUP iss	stern	Pup	GS	February	6
Pyxis	PIK sis	compass	Pyx	S	March	6
Reticulum	ruh TIK u lum	net	Ret	S	—	-
Sagitta	suh JIT uh	arrow	Sge	AG	September	17
Sagittarius	SAJ ih TAR ee us	archer	Sgr	ABGSZ	August	15
Scorpius	SKOR pee us	scorpion	Sco	ABGSZ	July	14
Sculptor	SCULPT tor	sculptor's shop	Scl	S	October	20
Scutum	SKU tum	shield	Sct	G	August	15
Serpens	SIR pens	snake	Ser	A	July	11,15
Sextans	SEX tans	sextant	Sex	S	—	5,6,7,8
Taurus	TAW russ	bull	Tau	ABGZ	January	1
Telescopium	tel uh SCO pee um	telescope	Tel	S	—	-
Triangulum	tri AN gue lum	triangle	Tri		November	22
Triangulum Australe	T. aw STRAY lee	s. triangle	TrA	S	—	-
Tucana	too CAY nuh	toucan	Tuc	S	—	-
Ursa Major	OOR suh MAY jor	great bear	UMa	ABC	April	9
Ursa Minor	OOR suh MY nor	lesser bear	Umi	AC	June	9
Vela	VEE luh	sails	Vel	BS	March	6
Virgo	VUR go	corn maiden	Vir	AZ	May	10
Volans	VO lans	flying fish	Vol	S	—	-
Vulpecula	vul PEK u luh	fox	Vul		September	17

Notes:
A = ancient
B = bright
C = circumpolar
G = galactic
S = southern
Z = zodiacal

Appendix

Which Sky Map to Use

	6 P.M.	7 P.M.	8 P.M.	9 P.M.	10 P.M.	11 P.M.	MIDNIGHT	1 A.M.	2 A.M.	3 A.M.	4 A.M.	5 A.M.	6 A.M.
Jan. 7	Nov		Dec		Jan		Feb		Mar		Apr		May
Jan. 23		Dec		Jan		Feb		Mar		Apr		May	
Feb. 7	Dec		Jan		Feb		Mar		Apr		May		Jun
Feb. 21		Jan		Feb		Mar		Apr		May		Jun	
Mar. 7	Jan		Feb		Mar		Apr		May		Jun		Jul
Mar. 23		Feb		Mar		Apr		May		Jun		Jul	
Apr. 7		Feb		Mar		Apr		May		Jun		Jul	
Apr. 22	Feb		Mar		Apr		May		Jun		Jul		Aug
May 7		Mar		Apr		May		Jun		Jul		Aug	
May 23	Mar		Apr		May		Jun		Jul		Aug		Sep
Jun. 7		Apr		May		Jun		Jul		Aug		Sep	
Jun. 23	Apr		May		Jun		Jul		Aug		Sep		Oct
Jul. 7		May		Jun		Jul		Aug		Sep		Oct	
Jul. 23	May		Jun		Jul		Aug		Sep		Oct		Nov
Aug. 7		Jun		Jul		Aug		Sep		Oct		Nov	
Aug. 23	Jun		Jul		Aug		Sep		Oct		Nov		Dec
Sep. 7		Jul		Aug		Sep		Oct		Nov		Dec	
Sep. 22	Jul		Aug		Sep		Oct		Nov		Dec		Jan
Oct. 7		Aug		Sep		Oct		Nov		Dec		Jan	
Oct. 23	Aug		Sep		Oct		Nov		Dec		Jan		Feb
Nov. 7	Sep		Oct		Nov		Dec		Jan		Feb		Mar
Nov. 22		Oct		Nov		Dec		Jan		Feb		Mar	
Dec. 7	Oct		Nov		Dec		Jan		Feb		Mar		Apr
Dec. 23		Nov		Dec		Jan		Feb		Mar		Apr	

Magnitude Scale for Sky Maps

5	fifth
4	fourth
3	third
2	second
1	first
0	zero
-1	Sirius

Key to Symbols

- Double and multiple stars
- Variable stars
- Globular clusters
- Planetary nebulae
- Supernova remnants
- Emission nebulae
- Open clusters
- Galaxies

Which Star Chart to Use

	6 P.M.	7 P.M.	8 P.M.	9 P.M.	10 P.M.	11 P.M.	MIDNIGHT	1 A.M.	2 A.M.	3 A.M.	4 A.M.	5 A.M.	6 A.M.
Jan. 7	21	22	23		1	2,3	4		5,6		7,8		9
Jan. 23	22	23		1	2,3	4		5,6		7,8		9	10
Feb. 7	23		1	2,3	4		5,6		7,8		9	10,11	
Feb. 21		1	2,3	4		5,6		7,8		9	10,11	12	
Mar. 7	1	2,3	4		5,6		7,8		9	10,11	12		
Mar. 23	3	4		5,6		7,8		9	10.11	12		13,14	15
Apr. 7	3	4		5,6		7,8		9	10,11	12		13,14	15
Apr. 22	4		5,6		7,8		9	10,11	12		13,14	15	16
May 7		5,6		7,8		9	10,11	12		13,14	15	16	17
May 23	6		7,8		9	10,11	12		13,14	15	16	17,18	
Jun. 7		7,8		9	10,11	12		13,14	15	16	17,18		
Jun. 22	7,8		9	10,11	12		13,14	15	16	17,18			19
Jul. 7		9	10,11	12		13,14	15	16	17,18			19,20	
Jul. 23	9	10,11	12		13,14	15	16	17,18			19,20		21
Aug. 7	11	12		13,14	15	16	17,18			19,20		21	22
Aug. 23	12		13,14	15	16	17,18			19,20		21	22	23
Sep. 7		13,14	15	16	17,18			19,20		21	22	23	
Sep. 22	13,14	15	16	17,18			19,20		21	22	23		1
Oct. 7	15	16	17,18			19,20		21	22	23		1	2,3
Oct. 23	16	17,18			19,20		21	22	23		1	2,3	4
Nov. 7			19,20		21	22	23		1	2,3	4		5,6
Nov. 22		19,20		21	22	23		1	2,3	4		5,6	
Dec. 7	19,20		21	22	23		1	2,3	4		5,6		7
Dec. 23		21	22	23		1	2,3	4		5,6		7,8	

The Greek Letters

α	alpha	η	eta	ν	nu	τ	tau
β	beta	θ	theta	ξ	xi	υ	upsilon
γ	gamma	ι	iota	ο	omicron	ϕ	phi
δ	delta	κ	kappa	π	pi	χ	chi
ϵ	epsilon	λ	lambda	ρ	rho	ψ	psi
ζ	zeta	μ	mu	σ	sigma	ω	omega

Magnitude Scale for Star Charts

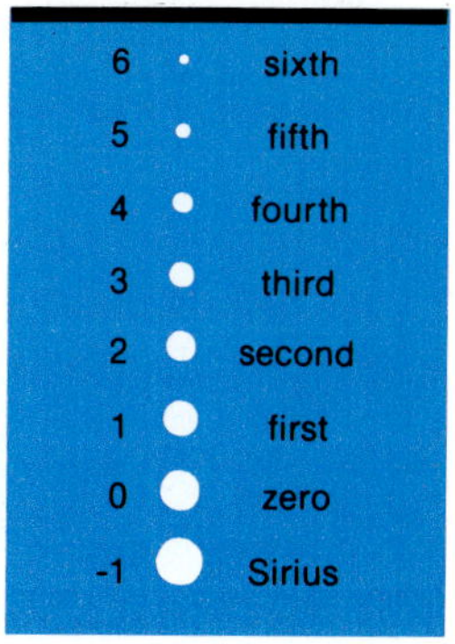

6	sixth
5	fifth
4	fourth
3	third
2	second
1	first
0	zero
-1	Sirius